INDUSTRY STANDARD OF THE PEOPLE'S REPUBLIC OF CHINA

Code for Design of Railway Station and Terminal

TB 10099-2017

Prepared by: China Railway Siyuan Survey and Design Group Co.,Ltd
Approved by: National Railway Administration
Effective date: December 1,2017

China Railway Publishing House

Beijing 2018

图书在版编目(CIP)数据

铁路车站及枢纽设计规范:TB 10099-2017:英文/中华人民共和国国家铁路局组织编译.—北京:中国铁道出版社,2018.11
ISBN 978-7-113-24586-3

Ⅰ.①铁… Ⅱ.①中… Ⅲ.①铁路车站-建筑设计-设计规范-英文 ②铁路枢纽-建筑设计-设计规范-英文 Ⅳ.①TU248.1-65

中国版本图书馆 CIP 数据核字(2018)第 123279 号

Chinese version first published in the People's Republic of China in 2017
English version first published in the People's Republic of China in 2018
by China Railway Publishing House
No. 8, You'anmen West Street, Xicheng District
Beijing, 100054
www. tdpress. com

Printed in China by BEIJING HUCAIS CULTURE COMMUNICATION CO., LTD.

© 2017 by National Railway Administration of the People's Republic of China

All rights reserved. No part of this publication may be reproduced or transmitted in any form or by any means, electronic or mechanical, including photocopying, recording, or by any information storage and retrieval systems, without the prior written consent of the publisher.

This book is sold subject to the condition that it shall not, by way of trade or otherwise, be lent, resold, hired out or otherwise circulated without the publisher's prior consent in any form of binding or cover other than that in which it is published and without a similar condition including this condition being imposed on the subsequent purchaser.

ISBN 978-7-113-24586-3

Introduction to the English Version

The translation of this Code was made according to Railway Engineering and Construction Development Plan of the Year 2016 (Document GTKFH [2016] 29) issued by National Railway Administration for the purpose of promoting railway technological exchange and cooperation between China and the rest of the world.

This is the official English language version of TB 10099-2017. In case of discrepancies between the original Chinese version and the English translation, the Chinese version shall prevail.

Planning and Standard Research Institute of National Railway Administration is in charge of the management of the English translation of railway industry standard, and China Railway Economic and Planning Research Institute Co., Ltd. undertakes the translation work. Beijing Times Grand Languages International Translation and Interpretation Co., Ltd. and China Railway Engineering Consulting Group Co., Ltd. provided great support during translation and review of this English version.

Your comments are invited and should be addressed to China Railway Economic and Planning Research Institute Co., Ltd., 29B, Beifengwo Road, Haidian District, Beijing, 100038 and Planning and Standard Research Institute of National Railway Administration, Building B, No. 1 Guanglian Road, Xicheng District, Beijing, 100055.

Email: jishubiaozhunsuo@126.com

The translation was performed by Jiang Hanke, Cheng Weizhou, Chai Guanhua, Dong Xuewu, Zhang Lixin, Liang Chao, Wang Xueyuan, Yue Ling.

The translation was reviewed by Chen Shibai, Wang Lei, Yang Quanliang, Liu Dalei, Geng Xin, Yu Xing, Zhu Zhexun, Xie Qiankun.

Notice of National Railway Administration on Issuing the English Version of Six Railway Standards including *Code for Design of Railway Alignment*

Document GTKF [2018] No. 83

The English version of *Code for Design of Railway Alignment* (TB 10098-2017), *Code for Design of Railway Station and Terminal* (TB 10099-2017), *Code for Design of Railway Track* (TB 10082-2017), *Code for Design of Steel Structure of Railway Bridge* (TB 10091-2017), *Code for Design of Subsoil and Foundation for Railway Bridge and Culvert* (TB 10093-2017) and *Technical Specification for Shield Tunnelling for Railway* (TB 10181-2017) is hereby issued. In case of discrepancies between the original Chinese version and the English version, the Chinese version shall prevail.

China Railway Publishing House is authorized to publish the English version.

National Railway Administration
October 18, 2018

Notice of National Railway Administration on Issuing Railway Industry Standard (Engineering and Construction Standard Batch No. 7, 2017)

Document GTKF [2017] No. 62

Code for Design of Railway Alignment (TB 10098-2017) and *Code for Design of Railway Station and Terminal* (TB 10099-2017) are hereby issued and will come into effect on December 1, 2017.

China Railway Publishing House is authorized to publish these Codes.

National Railway Administration
September 18, 2017

Foreword

This Code is prepared on the basis of comprehensively summarizing the practical experiences and scientific achievements on railway station and terminal design, construction and operation of China's high-speed railway, intercity railway, mixed traffic railway, and heavy-haul railway.

This Code implements the national development concept of "innovation, coordination, green, openness and sharing", and conforms to the policies and regulations on integrated traffic and transportation development, on natural ecology and environment protection, on land and energy saving, etc. In this Code, safety priority principle is highly strengthened; the idea of systematic design for stations and terminals, and the idea of capacity coordination between lines and stations are highlighted. The design principles and design parameters for stations and terminals are reasonably determined based on national conditions, economic and social development level, transportation demands and environment conditions, which greatly improves the scientificity as well as the technical and economic reasonability of this Code.

This Code consists of sixteen chapters: General Provisions; Terms; Basic Requirements; Terminal; Marshalling Station; District Station; Intermediate Station and Combination & Disassembly Station; Passing Station and Overtaking Station; Passenger Station, Passenger Facilities and Equipment and Passenger Car Depot (Shed); Railway Logistics Center; Hump; Industrial Station and Harbour Station; Border Station; Freight Collection Station and Freight Distribution Station; Earthworks and Drainage of Railway Station and Yard; Station Tracks. Also there are two Appendixes.

The main revisions are as follows:

1. The application scope is revised; the design speed of passenger train on mixed traffic railway is changed from 160 km/h to 200 km/h; contents about station and terminal designs are added for high-speed railway, intercity railway and heavy-haul railway.

2. Two chapters are added, i.e. "Border Station" and "Freight Collection Station and Freight Distribution Station", and the layout plan and main equipment of the two kinds of stations are included chiefly.

3. The section "Earthworks and Drainage of Railway Station and Yard" in the chapter "Basic Regulations of Station Design" in the previous edition is revised as an individual chapter "Earthworks and Drainage of Railway Station and Yard" in this Code. The following provisions are added: the classification of earthworks in station and yard, the design flood frequency and recurrence interval for earthworks in station and yard, the criteria of fill material and compaction for passenger platform, the arrangement of earthworks transition section, the design of water drainage facilities in severe cold areas, and so on.

4. The main principles for station design are proposed: people first, service to transportation, system optimization, and focusing on development. The main design contents of station and yard are specified: properly selecting the design standards and reasonably determining the station and

terminal design plan. The design requirements on integrality and systematicness for railway station and terminal are enhanced.

5. The following provisions are revised: the distance from track centerline to main building (structure) or to main equipment, the distance between centers of station tracks, and the minimum length of the tangent section between turnout end and curve starting/ending point. The following design provisions of route plans and profiles are integrated: the in-station connecting line, the connecting line between yards, the running track outside of the depot (operation point or work area), and the line inside of the depot (operation point or work area). The design provisions about installing the rail weighing bridge as well as installing the overload and unbalanced-load detection device on goods track are added. The provisions about installing the isolation device at selected stations are deleted for simultaneously receiving or receiving/departing the passenger car and freight car; and the provisions about how to arrange the main production buildings and office buildings in the station are deleted.

6. Relevant provisions are added about the planning of general view of railway terminal, the site selection and comprehensive development of passenger station and railway logistics center, the maintenance base of high-power locomotive, and EMU equipment.

7. The provisions are deleted about recommending the use of unidirectional layout plan in the new marshalling station; the provisions are canceled about ice adding operation, fish fry and livestock transportation, as well as the caboose.

8. Longitudinal-type layout plan of district station is added for passenger and freight transportation; longitudinal-type layout plan of district station for double-track railway is deleted; and the provisions is added that railway logistics center should be arranged at the same side as the shunting yard of district section.

9. The layout types and relevant provisions of combination and disassembly station are added; the layout types of intermediate station, crossing station and overtaking station are revised for mixed traffic railway.

10. The principle requirements of general design of passenger station are added; the layout types of stations of mixed traffic railway are revised; the provisions for passenger facilities and equipment in station are comprehensively integrated and revised; and the provisions of platform doors are added.

11. The provisions on the railway logistics center are newly added, including the general design principle, classification, function and military transportation requirements; the provisions for transporting the car-load, less-than-carload freight and ice adding for perishable freight are canceled; and provisions on relevant parameters of freight platform and warehouse are revised.

12. The following relevant provisions are deleted: using the steam locomotive for hump shunting locomotive, using the brake shoe and brake shoe taking-off device for hump speed regulator, and using the simple modern or manual speed regulators for small capacity hump.

13. Rail classification of station track is revised, all provisions about wooden sleeper are deleted, and the provision about the length of steel rail inserted into the adjacent single turnouts is revised.

We would be grateful if anyone finding the inaccuracy or ambiguity while using this Code would inform us and address the comments to China Railway Siyuan Survey and Design Group Co.,

Ltd. (No. 745 Heping Avenue, Wuchang, Wuhan, Hubei Province 430063, China) and China Railway Economic and Planning Research Institute (No. 29B, Beifengwo Road, Haidian District, Beijing 100038, China) for the reference of future revisions.

The Technology and Legislation Department of National Railway Administration is responsible for the interpretation of this Code.

Chief Technical Leader:

Zheng Jian, An Guodong, Wu Kefei.

Prepared mainly by:

China Railway Siyuan Survey and Design Group Co., Ltd.

And also by:

China Railway First Survey and Design Institute Group Co., Ltd.

China Railway Eryuan Engineering Group Co., Ltd.

China Railway Design Corporation

China Railway Engineering Consulting Group Co., Ltd.

Beijing National Railway Research & Design Institute of Signal & Communication Co., Ltd.

Drafted by:

He Zhigong, Liu Yiping, Han Guoxing, Wang Haichao, Lv Guojin, Yan Juping, Peng Jingping, Wang Huacheng, Zheng Hong, Li Mingguo, Ye Wenzhuo, Li Junying, Wan Fuying, Jin Guangrong, Zhou Binghe, Liu Shifeng, Lei Zhonglin, Gao Lan, Hu Mingxing, Shi Jianwen, Li Wenxiang, Sun Guocheng, Zhu Changqing, Li Changhuai, Li Shengcai, Li Chuanyong, Xiong Yuchun, Zhou Qinlong, Dong Zhiqiang, Cui Yongming, Dong Rukai, Meng Xianglei, Yao Chufeng, Xi Wenyuan, Zhang Kaizhi, Chen Chen.

Reviewed by:

Wu Kefei, Ding Liang, Wang Junfeng, Zeng Huixin, Xue Xingong, Xu Heshou, Ao Yunbi, Liu Yan, Sang Cuijiang, Wang Qiming, Wang Zhehao, Zhang Zhifang, Wan Jian, Ning Pei, Tian Yang, Wang Nan, Li Jianxin, Liu Hua, Zhang Yingfeng, Liu Shihui, Gao Fengnong, Kang Xuedong, Li Ronghua, Liu Xiaoping, Wu Wenxian, Liu Mingjun, Yang Jian, Wu Lirong, Jin Zude, Tian Changhai, Zhang Shihong, Li Shude, Nie Hongwang, Wu Maikui, Xu Youquan.

Contents

1 General Provisions ··· 1
2 Terms ·· 3
3 Basic Requirements ··· 5
   3.1 General Requirements ··· 5
   3.2 Plan of Station Approach Line and Station Track ································· 11
   3.3 Profile of Station Approach Line and Station Track ······························ 14
   3.4 Interface Design ·· 18
4 Terminal ·· 19
   4.1 General Requirements ··· 19
   4.2 Main Facilities and Equipment ··· 20
   4.3 Station Approach Line Layout and Untwining ··· 24
   4.4 Bypassing Line and Connecting Line ·· 25
5 Marshalling Station ·· 27
   5.1 General Requirements ··· 27
   5.2 Layout Plan of Marshalling Station ··· 27
   5.3 Main Facilities and Equipment ··· 29
   5.4 Number and Effective Length of Station Tracks ···································· 31
6 District Station ··· 34
   6.1 Layout Plan of District Station ·· 34
   6.2 Main Facilities and Equipment ··· 35
   6.3 Number and Effective Length of Station Tracks ···································· 36
7 Intermediate Station and Combination & Disassembly Station ························· 38
   7.1 Intermediate Station ·· 38
   7.2 Combination and Disassembly Station ·· 41
8 Passing Station and Overtaking Station ··· 43
   8.1 Passing Station ·· 43
   8.2 Overtaking Station ·· 43
9 Passenger Station, Passenger Facilities and Equipment and Passenger Car Depot(Shed) ··· 45
   9.1 General Requirements ··· 45
   9.2 Layout Plan of Passenger Station ·· 47
   9.3 Passenger Facilities and Equipment ·· 49
   9.4 Passenger Car Depot(Shed) ··· 52
10 Railway Logistics Center ··· 54
   10.1 General Requirements ··· 54

10.2	General Layout	54
10.3	Operation Car Yard	56
10.4	Freight Traffic Facilities and Equipment	56
11	Hump	59
11.1	General Requirements	59
11.2	Plan of Hump Track	59
11.3	Profile of Hump Track	60
11.4	Other Requirements	61
12	Industrial Station and Harbour Station	63
12.1	General Requirements	63
12.2	Layout Plan of Industrial Station and Harbour Station	63
12.3	Main Facilities and Equipment	65
12.4	Number and Effective Length of Station Tracks	66
13	Border Station	68
13.1	General Requirements	68
13.2	Layout Plan of Border Station	68
13.3	Main Facilities and Equipment	69
14	Freight Collection Station and Freight Distribution Station	71
14.1	General Requirements	71
14.2	Layout Plan of Freight Collection Station and Freight Distribution Station	71
14.3	Main Facilities and Equipment, Number and Effective Length of Station Tracks	74
15	Earthworks and Drainage of Railway Station and Yard	75
15.1	Earthworks in Railway Station and Yard	75
15.2	Drainage of Railway Station and Yard	77
16	Station Tracks	79
16.1	Track Design Standard	79
16.2	Rail and Auxiliary Parts	79
16.3	Sleeper and Fastenings	79
16.4	Track Bed	81
16.5	Turnout	82
Appendix A	Structure Gauge of Railway	84
Appendix B	Widening of Structure Gauge for Curve Section	89
Words Used for Different Degrees of Strictness		91

1　General Provisions

1.0.1　This Code is prepared with the view to unifying the technical standards of railway station and terminal design, and making the design meet the requirements on safety, reliability, advanced technology, economical efficiency and applicability.

1.0.2　This Code is applicable to the station and terminal design of standard-gauge high-speed railway, intercity railway, class Ⅰ and class Ⅱ mixed traffic railway, as well as heavy-haul railway.

1.0.3　In the process of railway station and terminal design, the following principles shall be complied with: people first, service to transportation, system optimization, looking forward to the future development, etc. The passenger and freight transport demands shall be analyzed systematically, the layout of facilities and equipment in the stations and terminals shall be coordinated as a whole, the design standards shall be selected properly and the design plans for stations and terminals shall be determined reasonably, according to the railway network planning and general urban planning as well as integrated transportation planning.

1.0.4　The design year for railway station and terminal shall be classified into short term and long term. The short term refers to the 10th year after being put into operation, and the long term refers to the 20th year after being put into operation. The predicted traffic volume shall be used for both the short term and long term traffic volumes. The design for the scales of railway infrastructures as well as equipment and buildings shall comply with following provisions:

1　The buildings and infrastructures which are hard to be renovated or expanded shall be designed according to long term traffic volumes and traffic types.

2　The buildings and infrastructures which are easy to be renovated or expanded shall be designed according to short term traffic volumes and traffic types, while long-term development conditions shall be reserved.

3　The operation equipment, whose allocation can increase or decrease with the change of transport demand, may be designed according to the predicted traffic volumes of the 5th year after being put into operation.

4　The development conditions for the far future shall be reserved in the general view of railway terminal.

1.0.5　The railway station and terminal design shall meet the system functional requirement and safety transportation requirement, shall facilitate passenger boarding and alighting as well as freight transporting, and shall facilitate both operation management and efficiency improving.

1.0.6　The railway station and terminal design plan shall be determined only after comprehensive technical economy comparison in accordance with general urban planning, existing railway conditions, topographical and geological conditions, local traffic conditions, farmland water conservancy facilities and land resource development, etc.

1.0.7　The railway station and terminal design shall meet the requirements of project construction by stages. The design of short-term project shall achieve the goals of reasonable layout, proper scale and convenient to operation, and also the further development conditions shall be reserved based on actual demand; the interference with railway operation caused by reconstruction operation and by abandoned works when the project reconstruction shall be reduced. The reconstruction and expansion projects shall take full advantage of the existing buildings, facilities and equipment; the design for guiding the construction transition shall be proposed in the case of complicated station reconstruction and expansion projects.

1.0.8 The railway station and terminal design shall pay attention to the overall coordination among different technical disciplines, and the designs shall be overall considered for station yard structures, cable trenches and troughs, water drainage facilities and lightning prevention and grounding.

1.0.9 Railway station and terminal design shall comply with the provisions of related law and regulations on environment, energy, land, cultural relic and firefighting.

1.0.10 Railway structure gauges shall comply with the provisions of Appendix A in this Code. The widening of structure gauges in curve sections shall comply with the provisions of Appendix B in this Code.

1.0.11 In addition to this Code, the railway station and terminal design shall also comply with the relevant national standards in force.

2 Terms

2.0.1 Railway station

The dividing point built for dealing with train passing-through, receiving and departure, technical operation, and passenger and freight transport services.

2.0.2 Railway terminal

A complete set of facilities built at the point or the end of railway network, which consists of at least two trunk lines, several train stations, various facilities serving the transport and their connecting lines.

2.0.3 Station approach line

The main line or terminal connecting line, led to each station in a railway terminal and built according to different train directions and train types.

2.0.4 Station track

A general term for all the tracks managed within a station, except for the main lines.

2.0.5 Receiving-departure track

The track built for train arrival and departure operation.

2.0.6 Other station tracks

A general term for all the tracks managed within the station, except for the main line and receiving-departure track as well as except for hump rolling track, which include shunting track, shunting neck, locomotive waiting track, running track, connection line in station, depot track, safety siding and so on.

2.0.7 Effective length of track

A certain track length within the range of the whole track length, on which the rolling stock can stay and not hinder the use of adjacent tracks.

2.0.8 Marshalling station

A station built for making up and breaking up of freight trains in large numbers.

2.0.9 District station

A station built for traction routing of the leading locomotive of freight trains, and also built for the breaking-up and marshalling operations of district trains and pick-up & drop-off trains.

2.0.10 Intermediate station

A station built for train passing through, train passing or crossing, train overtaking as well as passenger and freight transportation services.

2.0.11 Combination and disassembly station

A station built for meeting the needs of train combination and disassembly operation of heavy-haul railway.

2.0.12 Passing station

A station built on the single-track railway and used for train passing through, train passing or crossing and train overtaking in order to meet the needs of section carrying capacity.

2.0.13 Overtaking station

A station built on double-track railway and used for train overtaking in the same direction in order to meet the needs of section carrying capacity.

2.0.14 Passenger station

A station used mainly for dealing with passenger transportation service.

2.0.15 Passenger car depot (shed)

The place used specially for storage, servicing and maintenance of passenger train-set or EMU, including passenger car servicing depot, EMU running shed, EMU depot, etc. but not including the passenger transportation locomotive depot (shed).

2.0.16 Railway logistics center

A facility which relies on railway and provides logistics activities for the society with complete information network, featuring the functions such as providing logistics services for the society or for the enterprise itself, perfect logistics processes, large aggregation and radiation scopes, and strong storage and handling capacities, etc.

2.0.17 Hump

A shunting facility that is used for train breaking up, where the track in front of turnout area on the initial end of shunting yard is elevated to such a height that enables the car to roll automatically onto the shunting track by taking advantage of the potential energy and self weight of the car.

2.0.18 Industrial station

A station mainly undertaking the external railway transportation services for the enterprise with a great deal of loading/unloading operation.

2.0.19 Harbour station

A station mainly undertaking the external railway transportation services for the port with a great deal of loading/unloading operation.

2.0.20 Border station

A station built at the gateway of border, which is designated by the State for foreign contact and is a special node station for international passenger and freight transportation.

2.0.21 Freight collection station

A station used mainly for mass goods loading operation and for train set gathering operation.

2.0.22 Freight distribution station

A station used mainly for mass goods unloading operation and for train set disassembling operation.

3 Basic Requirements

3.1 General Requirements

3.1.1 At the tangent section of station tracks, the distance from the main building (structure) or equipment to the center line of the track shall comply with Table 3.1.1.

Table 3.1.1 Distance from Main Building (Structure) or Facility to the Track Centerline (mm)

No.	Building (structure) or facility description			High-speed railway	Intercity railway	Mixed traffic railway as well as heavy-haul railway	
						Height over rail surface	Distance to track centerline
1	Edge of pole or column like overpass bridge column, overbridge column, platform shelter pole and electric lighting pole	At main line side in station		≥2 440	≥2 200	—	≥2 440
		Between station tracks	With the passing of out-of-gauge freight train	—	—	1 100 and above	≥2 440
			No passing of out-of-gauge freight train	≥2 150	≥2 150	1 100 and above	≥2 150
		Outside of outmost station track of station yard		≥3 100	≥3 100	1 100 and above	≥3 100
		At the side of the outmost ladder track or shunting neck		≥3 100	≥3 100	1 100 and above	≥3 500
2	Edge of overhead contact system (OCS) pole	At the side of main line in station or outside of the outmost track of the station yard	Ballastless	≥3 000	≥2 500	—	—
			Ballasted	≥3 100	≥3 100	—	≥3 100
		Between station tracks	With the passing of out-of-gauge freight train	—	—	1 100 and above	≥2 440
			No passing of out-of-gauge freight train	≥2 150	≥2 150	1 100 and above	≥2 150
		At the side of the outmost ladder track or shunting neck		≥3 100	≥3 100	1 100 and above	≥3 500
3	Edge of high signal	High-speed railway and intercity railway	Main line	≥2 440	≥2 200	—	—
			Receiving-departure track	≥2 150	≥2 150	—	—
		Mixed traffic railway as well as heavy-haul railway	With the passing of out-of-gauge freight train	—	—	1 100 and above	≥2 440
			No passing of out-of-gauge freight train	—	—	1 100 and above	≥2 150
4	Edge of freight platform	Normal platform		—	—	950~1 100	1 750
		Elevated platform		—	—	≤4 800	1 850
5	Edge of passenger platform	Elevated platform	At the side of main line	1 800	1 800	—	—
			At the side of station track	1 750	1 750	1 250	1 750
		Common platform	At the side of receiving-departure track no passing of out-of-gauge freight train	—	—	500	1 750
		Low platform	At the side of receiving-departure track with the passing of out-of-gauge freight train	—	—	300	1 750

Table 3.1.1 (continued)

No.	Building (structure) or facility description		High-speed railway	Intercity railway	Mixed traffic railway as well as heavy-haul railway	
					Height over rail surface	Distance to track centerline
6	Edge of car shed door, car turning jack, car washing frame, tank washing siding as well as the building (structure) on locomotive running track		—	—	1 250 and above	≥2 000
7	Edge of dustman or switchman's cabin and enclosing wall		≥3 500	≥3 500	1 100 and above	≥3 500
8	From the edge of the fixing pole of hoisting device or from the edge of auxiliary equipment in running part to the goods loading/unloading siding		—	—	1 100 and above	≥2 440
9	Edge of diaphragm wall, fence and sound barrier	Located outside of main line or station track (without people passing)	Outside of earthworks surface	Outside of earthworks surface	—	Outside of earthworks surface

Notes: 1 In the case of ballasted track in No. 1 and No. 2 with large track maintenance machine operation, the distance from the inside edge of the pole in earthworks section to the centerline of main line shall not be less than 3 100 mm.
2 In the case of distance from the inside edge of OCS pole in No. 2 to the center line of track, if difficult shall not be less than 2 500 mm when located at the same side of ballasted main line, and shall not be less than 2 150 mm when located at the same side of the station track no passing of out-of-gauge freight trains.
3 In the case of No. 5, when the main line is without train passing or the speed of the passing train is less than or equal to 80 km/h, the distance from the edge of high platform to the center line of track may be 1 750 mm.
4 The distance from the edge of fence in No. 9 to the line center shall not be less than the sum of the height between fence and ground surface plus 2 440 mm for high-speed railway, and shall not be less than the sum of the height between fence and ground surface plus 2 200 mm for intercity railways.

3.1.2 The distance between center lines of two adjacent tracks in the station (or track spacing) shall meet the requirements of the operation and equipment arrangement between tracks, and shall be calculated and determined in accordance with the distance from the building (structure) edge or equipment edge to the line center, as shown in Table 3.1.1. Generally, the minimum distance between the tangent sections of the station tracks shall comply with the provisions specified in Table 3.1.2.

Table 3.1.2 Distance between Track Centerlines (or Track Spacing) in Station (mm)

No.	Item				Minimum distance between track centerlines
1	Between main lines in station	High-speed railway and intercity railway	Without crossover between main lines in station		The same as that of main lines in section
			With crossover between main lines in station	$v \leqslant 250$ km/h	4 600
				250 km/h$<v \leqslant$300 km/h	4 800
				300 km/h$<v \leqslant$350 km/h	5 000
		Mixed traffic railway			5 000
		Between the doubled-track and the third track, or between main lines in the same running direction			5 300
2	Between the main line in station and the adjacent receiving-departure track	Without the operation of train inspection, water supply and sewage discharging			5 000
		With the operation of train inspection, water supply and sewage discharging	$v \leqslant 120$ km/h	Normal	5 500
				Very difficult for reconstruction	5 000 (retained)
			120 km/h$<v \leqslant$160 km/h	Normal	6 000
				Very difficult for reconstruction	5 500 (retained)
			$v>160$ km/h	Normal	6 500 (with fence)
				Very difficult for reconstruction	5 500 (retained)

Table 3.1.2 (continued)

No.	Item		Minimum distance between track centerlines
3	Between receicing-departure tracks, between shunting tracks	Normal	5 000
		With train inspection trolley passage way paved or with the operation of passenger train water supply and sewage discharging	5 500
		Very difficult for reconstruction	4 600(retained)
4	Between tracks installed with high signals	Two adjacent tracks both with the passing of out-of-gauge freight trains	5 300
		Only one track of the two adjacent tracks may be with the passing of out-of-gauge freight trains	5 000
5	Between EMU stabling sidings		4 600
6	Between stabling sidings of passenger train-set	Normal	5 000
		Very difficult for reconstruction	4 600
7	Between servicing sidings of EMU and passenger trains	No electric pole for lighting or for communication between tracks	6 000
		With electric pole for lighting or for communication between tracks	7 000
8	Between tracks for direct transshipment of goods		3 600
9	Between shunting neck and adjacent track	District station, marshalling station and other stations with frequent train shunting operation	6 500
		Intermediate station and other station only with car pick-up and drop-off, as well as with car taking-out and placing-in operations	5 000
10	Between track groups in shunting yard		6 500
11	Between track groups in shunting yard which is equipped with brakeman's cabin		7 000
12	Between ladder track and adjacent tracks		5 000

Notes: 1 For No. 1 listed in above table, the distance between the centers of tracks shall be calculated and determined if reverse departure signal is installed between the main lines of intercity railway.

2 For No. 2 listed in above table, in the district station with train inspection operation, if the section design speed is 120 km/h or above, the safety measures to guarantee train inspectors must be taken in operation.

3 For No. 3 listed in above table, the train inspection trolley passageway should not be set between the receiving-departure tracks where the out-of-gauge freight trains are allowed to pass through; if there is mobile trolley passageway between the receiving-departure tracks, the distance between centers of adjacent receiving-departure tracks shall not be less than 6 000 mm.

4 In large stations like district station, marshalling station and so on, the distance between track centerlines shall be coordinated with the maximum transverse span of light bridge, OCS head span or portal structure; generally; a distance between track centerlines of not less than 6 500 mm shall be set every 8 lines or every 40 m at most, which should be set between two train yards or between two track groups.

5 The lighting and communication poles shall be concentratedly set between the tracks with wider track spacing when located in large stations with many station tracks, and should be set outside of the station tracks when located in the intermediate station. Other poles or columns should not be set at the same track spacing with the high signal, and the lookout conditions of the high signal shall be guaranteed if it is really required to be set at the same track spacing.

3.1.3 At curve section of station tracks, the distance from different buildings (structures) or equipment to center line of track as well as the track spacing shall be widened according to the provisions of Appendix B of this Code. The height of the passenger platform located inside of the curve shall be reduced if the track is of outer rail superelevation, the reduction values shall be the sum of 0.6 multiplying by outer rail superelevation.

3.1.4 For electrified railways, the range of overhead contact line system (OCS) erected on the station tracks shall be determined according to the following provisions:

1 All the lines accessed by electric locomotives and EMU trains shall be installed with OCS. There shall be installed with OCS within the 100 to 200 m effective length range of departure operation end of both the departure track and the marshalling-departure track accessed by electric locomotives, as well as along the departure route.

    2 The shunting neck and the freight track where the shunting operation is performed by electric locomotive shall be installed with OCS; when disturbed by hoisting machine or by other devices, the OCS shall be erected over the track section outside of the disturbed range.

    3 For stations equipped with diesel shunting locomotives, the shunting neck and freight track may be without OCS.

    4 The OCS shall not be erected on the following tracks: shunting track in station, goods loading/unloading siding with large lifting device, rolling stock depot track, station repair track, diesel locomotive stabling siding and servicing siding, track within maintenance workshop, light-oil depot track, industrial siding for flammable and explosive materials and other tracks not suitable for electrification.

    5 For district station, marshalling station and other large station with different traction types, the range of OCS erection shall be reasonably determined.

3.1.5 The tracks crossed by the head span of single group of OCS shall not exceed 8 within station range. The arrangement of OCS poles shall be well-matched with other equipment layout and with the long-term development.

3.1.6 For the overpass bridge which crosses above the electrified railway station, the height from its beam bottom to the rail surface of the track under bridge shall, in tangent section, comply with the following provisions:

    1 In the case of main line in station of high-speed railway and intercity railway, the above-mentioned height shall not be less than the sum of 5 650 mm plus the structural height of the OCS.

    2 In the case of main line in station of mixed traffic railway with design speed more than 160 km/h, the above-mentioned height shall not be less than 7 500 mm, and shall not be less than 7 220 mm under difficult conditions. In the case of main line in station of mixed traffic railway with design speed of not more than 160 km/h, the above-mentioned height shall not be less than 6 550 mm, and shall not be less than 6 200 mm under difficult conditions.

    3 In the case of main line in station which is with the passing of double-deck containers, the above-mentioned height shall not be less than the sum of 6 860 mm plus the structural height of the OCS.

    4 In the case of the marshalling station, district station or other stations with lots of shunting operation, the above-mentioned height shall not be less than 6 550 mm, and shall not be less than 6 200 mm under difficult conditions; if sufficient basis is available, the above-mentioned height of the existing overpass bridge shall not be less than 5 800 mm under very difficult conditions.

    5 The above-mentioned height of a hump overpass bridge which crosses above the locomotive running track shall be 6 000 mm; and shall not be less than 5 800 mm under difficult conditions.

    6 In areas with altitude of 1 000 m or above, the above-mentioned height shall be further heightened according to the provisions of *Code for Design of Railway Traction Power Supply* (TB 10009).

    7 In the case of the curve section with outer rail superelevation, the above-mentioned height shall be further heightened based on calculation.

3.1.7 The effective length of receiving-departure track in station shall comply with the following provisions:

    1 The effective length of the receiving-departure track for passenger trains shall be calculated and determined according to train marshalling length and the requirements of train control system. The effective length of the receiving-departure track of through-type station shall be 650 m for high-speed railway, shall not be less than 400 m for intercity railway which uses CTCS-2 train control system and parks 8 marshalling EMUs, and shall be 650 m for mixed traffic railway which only parks passenger trains. The effective length of the receiving-departure track used in dead-end station or used for unidirectional train receiving-departure shall be calculated and determined according to train marshalling length and the requirements of train control system.

  2 The effective length of the receiving-departure track of freight trains shall be determined according to the requirement of transport capacity, locomotive type and the length of car set pulled, and shall be coordinated with the effective length of the receiving-departure track of adjacent railway.

  3 The effective length of the receiving-departure track of mixed traffic railway shall be selected from the series of 1 050 m, 850 m, 750 m or 650 m.

  4 The effective length of the receiving-departure track of new heavy-haul railway shall be 1 700 m if the traction tonnage is 10 000 t; shall be 2 800 m if the traction tonnage is 20 000 t. The effective length of the receiving-departure track of reconstructed heavy-haul railway shall be determined according to the parameters of the locomotives and rolling stock.

  5 The effective length of the receiving-departure track of station located in helper locomotive section or pusher gradient section shall be the sum of the stipulated effective length plus the length of the helper locomotive.

3.1.8 The main line in the station of mixed traffic railway shall be available for the passing of out-of-gauge freight trains.

The locomotive exchanging station as well as 3 to 5 passing stations, overtaking stations and intermediate stations selected in the district shall meet the requirements for passing and overtaking of out-of-gauge freight trains. In addition to the main line in above-mentioned stations, there shall be another one for the single-track railway, and also another one respectively for the up direction and the down direction of double-track railway, that can be available for the passing of out-of-gauge freight trains.

3.1.9 Track junction and safety siding shall comply with the following provisions:

  1 The junction of new track with the existing track shall ensure that there is no change of running direction when the main-direction train passing through the track junction point.

  2 The new tracks shall be connected inside the station. If they must be connected in section under difficult conditions, the block house shall be set at the track junction point, and the safety siding shall be set at the track junction point.

  3 The depot track as well as the branch track shall be connected in station, and shall not be connected with the main line in section. There may be without safety siding at the junction between station track and locomotive depot, at the junction between station track and passenger car serving shed, or at the junction between station track and EMU depot (shed). There may be without safety siding at the junction between station track and other depot tracks or at the junction between station track and branch tracks if there are other station track and the turnout that can be used as isolation facility with interlocking device; safety sidings shall be set for all the other cases.

  4 For mixed traffic railway and heavy-haul railway, if the equivalent gradient of arriving direction is greater than or equal to 6‰ down slope within the emergency braking distance outside of home signal, the safety siding shall be set at the tail end of station receiving track; safety siding may not be set if there are other station track or branch track at the tail end of station receiving track which can be used as isolation device.

5 Safety siding shall be set if there is no parallel route isolation at the junction point where the shunting route of the station is directly connected with the main line.

6 If there is the track which is directly connected to the main line in station or to the receiving-departure track and on which the passenger and freight trains or cars are parked for a long time, the safety siding shall be set at the junction point where the line is connected to the running route.

3.1.10 The design of safety siding shall comply with the following provisions:

1 The effective length of the safety siding shall not be less than 50 m.

2 The longitudinal grade of the safety siding shall be designed as zero slope or up slope towards the car bumper.

3 The tail part of the safety siding should not be set on bridge or in tunnel.

4 The safety siding shall be set with guard rail at both sides, its tail part shall be set with car bumper and buffer device, and the tail part of the safety siding on earthworks section shall be set with wheel-stopping soil structure.

3.1.11 The station equipped with shunting locomotive shall be furnished with shunting locomotive serving equipment at proper site.

3.1.12 The arrangement of level crossing in station shall comply with the following provisions:

1 Level crossing in station shall not be set for high-speed railway, intercity railway or mixed traffic railway with section design speed of more than 120 km/h, and grade separation shall be used for track crossing.

2 When level crossing is required to be set in a station, for the case of station with intermediate platform, one level crossing should be set at the middle part of the station to connect the intermediate platform and the main platform; for the case of station with passenger overpass, one level crossing shall be set at the middle part of the station based on actual requirement.

3 For a receiving yard, receiving-departure yard, departure yard or for a marshalling-departure track, where train inspection operation needs to be conducted, the level crossing may be set at the end of the train yard or outside of the fouling post.

4 Both ends of the car retarder of rolling section of hump, in front of the small capacity hump track-group turnout and in front of the car retarder of shunting track, the level crossing may be set at proper site based on road layout in the station.

5 The level crossing shall be set at both ends or at one end of the stabling siding or servicing siding of passenger car depot (shed). When two level crossings need to be set, their spacing shall not be less than the total length of the EMU or passenger train-set, and shall not be less than the sum of above-mentioned total length plus 10 m for the uncoupled-car inspection if set at the technical servicing siding.

6 The level crossing may be set for other yard, depot or operation point according to actual requirement.

7 The width of the level crossing in station shall be determined according to its purpose. It may be 1.5 m wide if only for the walking of station staff; it may be 2.5 m wide if for the passing of non-motor vehicle; and shall not be less than 3.5 m wide if for the passing of motor vehicle.

8 The level crossing inside the station and yard shall be separated from the road and sidewalk outside of the station, which shall be used only by station staff or vehicles.

3.1.13 Road system shall be set in stations; the district station, marshalling station and other large stations shall be provided with the circular road which shall surround the car yard and shall be connected with the local roads. For the overpass bridge crossing the main roads in the station, its clearance shall meet the requirement of firefighting and the passage of transport vehicle.

3.1.14 If the section design speed is 120 km/h or above for the main line, the station shall be set with isolation devices like protective fence or enclosing wall, which shall be connected with the section isolation device.

3.1.15 The numbering of station tracks shall comply with the following provisions:

  1 For the station located in single-track district, the tracks shall be numbered sequentially from the track near the station building to the track far away from the station building.

  2 For the overtaking station and intermediate station located in double-track district, the tracks shall be numbered sequentially from the main line, and there shall be even numbers for up direction tracks and odd numbers for down direction tracks.

  3 For dead-end stations, the tracks shall be numbered sequentially towards the destination direction from the left side; if the station building is located at one side of the tracks, the tracks shall be numbered sequentially from the track near the station building to the track far away from the station building.

  4 The tracks within district station, large and super large passenger stations shall be numbered sequentially with the main platform of main building as the benchmark; if there are many car yards, the tracks of different car yards shall be numbered continuously in order instead of car yard type numbering; the other tracks within the car yard shall be numbered after the numbering for main lines and for receiving-departure tracks, and the numbering shall be conducted in the sequence of down-direction end prior to up-direction end and from the side of the (main) station building to the opposite side sequentially.

  5 Except for being laid in the district stations as well as large and super large passenger stations, the tracks laid in the station with many car yards and other attached yards, shall be numbered according to different yards and shall be named after yard name or yard number.

  6 The receiving-departure track and other station track shall be numbered with Arabic numerals, and the main line shall be numbered with upper roman numerals.

3.2 Plan of Station Approach Line and Station Track

3.2.1 The plan of station approach line shall adopt the standard of main line of adjacent section. The minimum curve radius of the untwining track with passenger train passing through shall not be less than 400 m under difficult conditions, the minimum curve radius of other untwining tracks shall not be less than 300 m, and the minimum curve radius for both the receiving loop and departure loop in marshalling station as well as for the loading (unloading) loop shall not be less than 250 m.

3.2.2 The curve radius of the connecting track in station and the connecting track between yards shall not be less than 400 m with passage of passenger trains, shall not be less than 250 m with passage of common freight trains, and shall not be less than 200 m without train passage.

3.2.3 The car yard located in marshalling station or in combination and disassembly station shall be arranged at straight line; under difficult conditions, the receiving yard, departure yard as well as the receiving-departure yard shall be arranged at the curve in the same direction, and the curve radius shall not be less than 800 m.

3.2.4 The track beside the high passenger platform shall be arranged at straight line; under difficult conditions, it shall be arranged at curve with radius of not less than 1 000 m; under very difficult conditions, the curve radius shall not be less than 600 m.

3.2.5 The freight line plan shall comply with the following provisions:

1 Goods loading/unloading siding shall be arranged at straight line; under difficult conditions, it shall be arranged at the curve with radius of not less than 600 m, and under very difficult conditions, the curve radius shall not be less than 500 m.

2 The loading/unloading siding for tank car freight shall be of dead-end flat straight line, and the distance from the starting point of the straight line section of the track to the first loading arm of the loading trestle shall not be less than 1/2 of the tank car length; under difficult conditions, it shall be arranged at the curve with radius of not less than 600 m. The safety distance from the center line of car coupler at the starting end of the tank car on the loading/unloading siding to the turnout fouling post ahead shall not be less than 31 m, and the safety distance from the center line of the car coupler at the tail end to the stop buffer of the loading/unloading siding shall not be less than 20 m.

3 The plan of the track which is equipped with rail weighing bridge, freight train overload and unbalance-load detection device, quantitative loading system, car tippler, etc., shall be designed according to technical requirements of equipment.

3.2.6 The layout of shunting neck shall comply with the following provisions:

1 The shunting neck shall be arranged at straight line; under difficult conditions, it shall be arranged at the curve with radius of not less than 1 000 m; under very difficult conditions, the curve radius shall not be less than 600 m. For the shunting neck of the logistics center or other yard (depot) where only car pick-up and drop-off operation as well as car taking-out and placing-in operation are carried out, its curve radius shall not be less than 300 m under very difficult conditions.

2 The shunting neck shall not be arranged at reverse curve; in the case of reconstructed stations with less shunting operation under very difficult conditions, it may be arranged at the reverse curve, and the existing curve radius may be retained.

3.2.7 The plan of the running track outside of the depot (operation point or work area) shall comply with the following provisions:

1 The curve radius of the running track outside of the passenger car depot (shed) should not be less than 400 m, and shall not be less than 300 m under difficult conditions.

2 The turnaround track of EMU should be arranged at straight line, and under difficult conditions, it shall be arranged at the curve with radius of not less than 600 m; the radius of the track with which the turnaround track is connected with the station throat shall not be less than 300 m. The effective length of the turnaround track for EMU shall be calculated and determined according to the marshalling length of EMU, and the effective length of the turnaround track shall not be less than 480 m and 270 m if the EMU is marshaled by 16 cars and 8 cars respectively.

3 The curve radius of the running track outside of locomotive depot (shed), car depot (shed) and maintenance base (workshop, work area) shall not be less than 200 m.

4 If the running track outside of the depot (shed) also serves as shunting neck, the provisions of layout for the shunting neck shall also be complied with.

3.2.8 The line plan inside of depot (operation point or work area) shall comply with the following provisions:

1 The track inside of the depot (operation point or work area) should be arranged at straight line, and the track inside the workshop (shed) shall be arranged at straight line.

2 The minimum curve radius of the track inside of passenger car depot (shed) shall not be less than 250 m, and the minimum curve radius of the track shall not be less than 400 m where EMU is parked for a long time; the minimum curve radius of the track inside of other depot (shed) shall not be less than 200 m.

3 The length of straight line section in front of the turning jack in the locomotive depot shall not be less than 12.5 m long.

3.2.9 The layout of curve superelevation for station track shall comply with the following provisions:

1 The curve superelevation shall be calculated and determined according to the horizontal curve radius and train passing speed, and shall comply with the requirements of the allowable deficient superelevation, the allowable surplus superelevation as well as the sum of the deficient superelevation plus the surplus superelevation corresponding to the track property.

2 The curve shall be set with curve superelevation for receiving-departure route of EMU train, and shall not be less than 20 mm. The curve and connecting curve of other passenger and freight receiving-departure line should be set with curve superelevation, it may be 25 mm if the curve radius is less than 600 m, or 20 mm if the curve radius is 600 m or above; the superelevation of the connecting curve for turnout shall be 15 mm. Other station tracks may not be set with superelevation.

3 The superelevation runoff rate shall not be more than 2‰.

3.2.10 The curve of the receiving-departure route for EMU train shall be set with spiral curve. The length of the spiral curve shall be calculated and determined according to track property, train passing speed, curve design superelevation, the time-varying rate of superelevation or deficient superelevation, and superelevation runoff rate, and shall not be less than 20 m. If station throat adopts No. 18 turnout, and the curve radius is 1 200 m or above, the spiral curve may not be set; if station throat adopts No. 12 turnout and the curve radius is 400 m or above, the spiral curve may not be set. Other station tracks may not be set with spiral curves.

3.2.11 When the spiral curve is set for the curve of the passenger train receiving-departure route, the circular curve and the tangent between the two curves shall not be less than 25 m long. When the spiral curve is not set, the tangent section without superelevation between the two curves shall not be less than 20 m long, and shall not be less than 10 m long under difficult conditions if No. 12 turnout is adopted. The tangent section between two curves for other station tracks shall not be less than 15 m long, and shall not be less than 10 m long under difficult conditions.

3.2.12 The length of the tangent section between the turnout and the curve shall comply with the following provisions:

1 The minimum length of the tangent section between the turnout on the main line and the spiral curve shall be calculated and determined according to the railway nature, section design speed, engineering condition, station nature and train acceleration and deceleration property, and shall comply with the requirements specified in Table 3.2.12-1.

Table 3.2.12-1 Minimum Length of Tangent Section between Main Line Turnout and Spiral Curve

Railway nature		High-speed railway		Intercity railway		Mixed traffic railway as well as heavy-haul railway		
Design speed v (km/h)		$250 \leqslant v \leqslant 350$	200	<200	200	$120 < v < 200$		$\leqslant 120$
Minimum length of tangent section (m)	Normal	0.6v		0.4v		0.5v		
	Difficult	0.5v	30	25	0.4v	30		20

Note: Under difficult conditions, the calculated value shall be valued downward as per the multiple of 10 m.

2 The length of the tangent section between the turnout of station track and the curve shall be calculated and determined according to station track nature, curve radius, turnout structure, curve gauge widening and curve superelevation, etc.

1) The length of the tangent section shall not be less than 20 m from the front end or back end of the turnout on EMU train receiving-departure route to the tail end of curve

superelevation runoff; under difficult condition, the length of the tangent section in front of the turnout shall meet the requirement of curve superelevation, and the length of the tangent section in back of the turnout shall not be less than the sum of the distance between turnout heel end and the last long turnout sleeper plus the required length of superelevation runoff. The minimum length of the tangent section between the turnout and the curve inside the EMU depot (shed) shall not be less than 7.5 m in front of the turnout, and in back of the turnout, shall not be less than the sum of the distance between turnout heel end and the last long turnout sleeper plus the required length of the tangent section for curve gauge widening declining or curve superelevation runoff.

2) The minimum length of the tangent section between turnout and circular curve for other station tracks shall comply with those specified in Table 3.2.12-2.

Table 3.2.12-2 Minimum Length of Tangent Section between Turnout and Circular Curve

No.	Circular radius in front and in back of turnout R (m)	Minimum length of the tangent section(m)			
		Normal		Difficult	
		Track gauge widening or curve superelevation declining rate 2‰		Track gauge widening declining rate 3‰	
		In front of turnout	In back of turnout	In front of turnout	In back of turnout
1	R≥350	2	0+L'	0	0+L'
2	350>R≥300	2.5	2.5+L'	2	2+L'
3	R<300	7.5	7.5+L'	5	5+L'

Notes: 1 L' is the distance from turnout heel end to last turnout sleeper.
2 Under difficult conditions, the tangent length in back of turnout may utilize the calculated length which is obtained after L' specified in Table 3.2.12-2 is replaced by the distance between turnout heel end and the last long turnout sleeper.

3) When spiral curves are provided for the curves in front and in back of turnout, the tangent section may not be inserted.

4) The track turnout located in rolling section of hump may not be inserted with tangent section, but shall be directly connected with the circular curve instead.

3.2.13 The radius of the connecting curve in back of the turnout shall be coordinated with the lateral allowable passing speed of the adjacent turnout.

3.3 Profile of Station Approach Line and Station Track

3.3.1 The profile of station approach line shall comply with the following provisions:

1 The maximum gradient of the track for only EMU trains should not be more than 20‰, and shall not be more than 30‰ under difficult conditions.

2 The maximum gradient of the track for passenger trains with locomotive traction should not be more than 15‰, and shall not be more than 20‰ under difficult conditions.

3 The track only for the unidirectional freight train may be set on the down slope which exceeds the ruling gradient, and its maximum gradient: shall not be more than 12‰ for single locomotive traction, and shall not be more than 15‰ under very difficult conditions; when using pusher traction, the maximum gradient shall not be more than 30‰ for electric traction, and shall not be more than 25‰ for diesel traction. If the track is used for reverse running, the traction calculation shall be performed and the track shall meet the requirement for the train passing through the track with speed of not less than the calculated train speed. The gradient difference of the adjacent slope sections shall comply with those specified in Table 3.3.1.

Table 3.3.1 Maximum Gradient Difference of Adjacent Slope Sections for the Track with Passage of Freight Trains

Long-term effective length of receiving-departure track for freight trains(m)		1 050 and above	850	750	650
Maximum gradient difference(‰)	Normal	8	10	12	15
	Difficult	10	12	15	18

  4 The profiles of station approach tracks except the above mentioned cases shall adopt the standards of the main lines of the adjacent railway sections.

3.3.2 The profiles of different yards of technical operation station and related tracks shall comply with the following provisions:

  1 The receiving yard, receiving-departure yard and departure yard should be set on the zero slope, and shall be set on the slope of not more than 1‰ under difficult conditions.

  2 The profile of shunting yard shall be determined according to the adopted speed-control equipment and corresponding control mode and technical requirements.

  3 If the main line is used for uncoupled-car repairing in the receiving-departure yard, departure yard and transit yard, the profile of the main line shall meet the starting condition of half train during train shunting.

  4 If huge engineering quantity may be resulted from using the above mentioned standards for reconstructed receiving yard, receiving-departure yard, departure yard and transit yard, the existing gradient shall be retained with the approval of related authorities, but the anti-running safety measures shall be taken.

  5 The gradient of the connecting line between train yards shall meet the requirement for yard transition of the whole train set.

3.3.3 The shunting neck with breaking-up and marshalling operation should be set on the zero slope, or set on the down slope which is not more than 2.5‰ and faces the shunting track. The shunting neck with car pick-up and drop-off operations as well as car taking-out and placing-in operations in logistics center, or the shunting necks in other yards and depots should be set on the slope no more than 1‰; or set on the slope no more than 6‰ under difficult conditions.

3.3.4 Goods loading/unloading siding shall be set on the zero slope, and shall be set on the slope of not more than 1‰ under difficult conditions; the loading/unloading siding for liquid goods and hazardous goods shall be set on the zero slope. The distance shall not be less than 15 m from the origin/destination point of goods loading/unloading siding to the origin/destination point of convex vertical curve. The gradient of the track shall be designed according to technical requirements of the equipment, which is equipped with rail weighing bridge, overload and unbalance-load detection device for freight trains, quantitative loading system and car tippler, etc.

3.3.5 The gradient of the running track outside of depot (operation point or working area) shall comply with the following provisions:

  1 The maximum gradient of the running track for EMU depot (shed) should not be more than 30‰, and shall not be more than 35‰ under difficult conditions.

  2 The gradient of the taking-out and placing-in track for passenger train-set should not be more than 12‰, and shall be determined according to traction calculation under difficult conditions; its gradient shall not be more than 6‰ if it also serves as shunting neck.

  3 The gradient of the locomotive running track outside of depot (shed) should not be more than 12‰, and shall not be more than 30‰ if there is interchange. At the dividing point of station and depot (shed), the locomotive parking space shall be kept whose length shall be the length of 2 locomotives plus 10 m, and its gradient shall not be more than 2.5‰.

4 The gradient of car transfer track for depot shall meet the requirements for car taking-out and placing-in as well as the requirements for track transition and shunting in the depot.

5 The gradient of the running track for train maintenance and repairing should not be more than 30‰, and shall not be more than 35‰ under difficult conditions.

3.3.6 The gradient of track in depot (operation point or working area) shall comply with the following provisions:

1 The station repair track and tank washing siding as well as the track inside of building (structure) shall be set on the zero slope. Other tracks inside the depot (operation point or working area) should be set on the zero slope, and may be set on the slope of not more than 1‰ under difficult conditions.

2 The locomotive waiting track shall be set on the zero slope or set on the up slope of not more than 5‰ towards the car bumper. In front of the turning jack, the zero slope section greater than or equal to 50 m long shall be set.

3 The gradient of the track in throat area of depot (operation point or working area) should not be more than 2.5‰, and shall not be more than 6‰ under difficult conditions.

3.3.7 The length of slope section for station track shall comply with the following provisions:

1 The length of slope section for station approach line shall adopt the standard of the main line of the adjacent railway section, and the length of slope section of untwining line shall not be less than 200 m under difficult conditions.

2 The effective length range of the receiving-departure track for high-speed railway station and for intercity railway station should be designed as one slope section; under difficult conditions, the length of slope section of the receiving-departure track for high-speed railway shall not be less than 450 m, and the length of slope section of the receiving-departure track for intercity railway shall not be less than 250 m; the length of slope section of the receiving-departure track for mixed traffic railway station as well as for heavy-haul railway station should not be less than those specified in Table 3.3.7.

Table 3.3.7 Minimum Slope Section Length of Receiving-Departure Track in Station (m)

Long-term effective length of receiving-departure track	1 050 and above	850	750	650
Slope section length	400	350	300	250

Note: For railway section with the section design speed 160 km/h and above, the minimum slope section length shall not be less than 400 m, and should not be applied continuously for twice or more.

3 The slope section lengths of station tracks except the receiving-departure track shall not be less than 200 m with train passage, and shall not be less than 50 m without passage of trains.

3.3.8 The connection for slope sections of station track shall comply with the following provisions:

1 The connection for slope sections of station approach track shall adopt the standard of the main line of the adjacent line.

2 When the gradient difference of the adjacent slope sections of receiving-departure track of high-speed railway or intercity railway is more than 3‰, the vertical curve with radius of not less than 5 000 m shall be applied to connection.

3 For the receiving-departure tracks and the station tracks with train passage of mixed traffic railway as well as heavy-haul railway, if the gradient difference of adjacent slope sections is more than

4‰, the vertical curve with radius 5 000 m may be applied; under difficult conditions, the vertical curve radius shall not be less than 3 000 m.

4 For the station track without train passage, if the gradient difference of adjacent slope sections is more than 5‰, the vertical curve with radius of 3 000 m may be applied; under difficult conditions, the locomotive running track with interchange may adopt the vertical curve with radius of not less than 1 500 m; the elevated loading/unloading siding may adopt the vertical curve with radius of not less than 600 m.

3.3.9 The layout for the turnout, the vertical curve starting and finishing points, and the gradient change point shall comply with the following provisions:

1 For high-speed railway and intercity railway, the distance should not be less than 20 m from the two ends of main line turnout to the vertical curve starting/finishing point or to the gradient change point.

2 For mixed traffic railway and heavy-haul railway, the main line turnout shall not be set within the range of vertical curve or on the gradient change point.

3 The turnout of the receiving-departure track shall not be overlapped with the vertical curve or with the gradient change point.

4 The turnouts of other station tracks should not be overlapped with the vertical curves or with the gradient change points; under difficult conditions, if it has to be set for reconstructed railways, the vertical curve radius shall not be less than 10 000 m for the station track with train passage, and shall not be less than 5 000 m for the station track without train passage.

3.3.10 The rail surface height difference and gradient transition section of station tracks shall comply with the following provisions:

1 For high-speed railway station and intercity railway station, the rail surface from the main line in station to the receiving-departure track, from the receiving-departure track to the receiving-departure track should be designed as per equal altitude. If there is rail surface height difference between throat area tracks, the gradient transition section of rail surface height difference shall be designed according to the transverse gradient of earthworks surface and the thickness of track bed. The range of gradient transition section for receiving-departure track shall be from the common sleeper both ends of the turnout to the departure signal. The gradient of gradient transition section should not be more than 6‰, the algebraic difference of the gradients of adjacent slope sections should not be more than 3‰, and the length of slope section shall not be less than 50 m.

2 In mixed traffic railway station and heavy-haul railway station, if there is rail surface height difference between two adjacent tracks within the throat area, the gradient transition section shall be designed according to the ruling gradient of the main line, station and yard gradient, transverse gradient of earthworks surface as well as the thickness of track bed. The range of gradient transition section should be from the first common sleeper in back of turnout rear-end to the fouling post or to the effective length starting-point of goods loading/unloading siding. The gradient difference between adjacent slope sections of the gradient transition section should not be more than 4‰ for receiving-departure track and for station track with train passage, and should not be more than 5‰ for other station tracks; the length of slope section shall not be less than 50 m.

3 If the drop height does not meet the requirement of the gradient transition section, measures shall be taken for adjustment, based on the actual condition of the station, such as reducing the transverse gradient of earthworks surface, track bed thickening, paving double-layer track bed, or intruding the gradient transition section into the effective length of the track.

3.4 Interface Design

3.4.1 The columns, network and pipeline layout in the station and yard shall be systematically designed, overall considered and shall be coordinated with the station and yard layout.

3.4.2 The design width of station earthworks and the design width of section earthworks shall be smoothly connected with each other at their interface; the protection standard and greening standard of the station earthworks shall be coordinated with that of the section earthworks.

3.4.3 The cable trough between station and section as well as between earthworks and bridge/culvert shall be reasonably connected according to the technical requirements for the laying of cable trough.

3.4.4 The cable trench and trough, the facility for pipeline crossing the track, the inspection hole and other facilities shall be synchronously designed and constructed with the earthworks in the station and yard.

3.4.5 The width of the station earthworks shall meet the requirements for the layout of cable trench and trough as well as sound barrier.

3.4.6 The station platform surface, shelter, fence and so on located within the station and yard, whose foundation contains metal structure, shall be connected to the integrated through earthing wire according to related technical requirements.

3.4.7 The drainage interface design for station and yard shall comply with the following provisions:

1 The drainage system in station and yard shall be effectively linked with the section drainage facilities.

2 The drainage system of station and yard shall be integrally designed based on bridge and culvert layout, railway water drainage pipeline networks and urban water drainage system.

3 When drainage system of station and yard is connected to the bridge or culvert, the inlet elevation shall be higher than the elevation of the drainage outlet of the bridge or culvert.

4 If the poles of overhead contact system or columns of platform shelter in station are located between the tracks with drainage channel (ditch), the related pole or column foundations shall be designed integratedly with the drainage channel (ditch).

3.4.8 The turnout shall not be arranged at the junction where the embankment is connected with bridge abutment; the turnout of the main line should not be arranged at the transition section between earthworks and bridge, between earthworks and tunnel, or between bridge and tunnel.

3.4.9 The passageway of passenger entering and exiting station shall be synchronously designed and constructed with the station earthworks, and the location and elevation of the station access shall meet the technical requirements for the layout of drainage channel and cable trough in station.

4 Terminal

4.1 General Requirements

4.1.1 The planning of general view of terminal shall integrate the factors as follows: the coordination of station and line capacity, multi-network integration, passenger lines inside of urban area and freight lines outside of urban area, paying attention to both the passenger train and freight train operation, making full use of the existing facilities, overall planning for new line leading in and reconstruction of the exiting line, overall development for high-speed railway and conventional railway, overall planning for station and yard layout and auxiliary facilities. All the factors shall be on the basis of long term, according to the principle of long term and short term combination, as well as implementation by stages, which will embody the systematicness, perceptiveness and economy, and the terminal layout will be optimized and will be of perfect function.

4.1.2 The design of general view of terminal shall be comprehensively compared and selected based on the overall situation, urban planning and other traffic and transportation systems; the role and scale of the terminal, the technical feature of led in lines, the feature and flow of passenger and freight traffic, condition of existing equipment, topographical and geological conditions shall be integrally analyzed.

4.1.3 The terminal construction shall be implemented by stages according to the general view of the terminal, and land shall be reserved as per the requirement of long-range perspective. The short-term projects shall aim at reasonably layout, proper scale, convenient operation, effective investment cost and remarkable economic benefit, and the abandoned works resulted from reconstruction process shall be possibly reduced, and the interference of the construction with the operation shall be reduced either.

4.1.4 Based on local conditions, the general view of terminal shall be designed according to the following requirements:

1 When there are few lead-in lines, small volume of passenger and freight transport, and the city is small, one-stop terminal shared by passenger and freight may be designed.

2 If the lead-in lines converge from three directions, the passenger and freight shared station, other stations or block station may be established at the confluence to form the triangle-type junction terminal, based on the passenger and freight traffic volume of different directions.

3 If the newly constructed line with large traffic flow crosses the existing line, the necessary station and untwining line to connect the existing line may be constructed, rendering the newly-constructed line to directly cross the existing line to form cross-type junction terminal.

4 If there are many lead-in lines, large passenger and freight traffic volume, two or more special-purpose stations are required to be established based on urban planning and local conditions, the main passenger station and marshalling station may be designed as aligned or paralleling terminal. For aligned terminal, the lead-in lines from both ends may be properly dealt with, with attention to the passing capacity of intermediate busy section, and the bypassing line shall be set if necessary. If the city is divided into different sections by rivers, the main passenger and freight transportation facilities of the terminal shall be set at city side where the lead-in lines converges, and the passenger and freight transportation facilities may be arranged in each different areas if necessary.

5 If the linked lines are of different directions and located in the terminal of big city, they may be designed as circular or semi-circular terminal based on line direction, station distribution and the connecting line which provides service to cities and industrial parks, the loop line should be set outside of urban area which can provide flexible and convenient passageway for the lead-in lines from different directions. For circular terminal around megalopolis, direct lines may be built between stations if necessary. The loop line of super-large terminal may be designed as passenger loop line or freight loop line with passenger line inside of urban area and freight line outside of urban area.

6 For dead-end type junction terminal which is located in port city or mine area and at the end of railway network, the marshalling station should be set at the entrance or exit, which will facilitate the exchange of local trains from different areas.

7 If one type of terminal layout can not meet operation requirement, the compound terminal may be designed that is composed of several terminal types and is correspond with the terminal operation volume and operation nature.

4.1.5 The new lines which are led in the terminal should not be directly connected with the marshalling station excessively; under normal circumstances, they may be connected at the station in front of the terminal or the appropriate station in the terminal.

4.1.6 The new industrial siding group with certain scale in the terminal shall be overall planned, and the track connection stations shall be reasonably selected, on the basis of terminal layout, industrial area distribution and urban construction.

4.2 Main Facilities and Equipment

I Passenger Station

4.2.1 The number and configuration of passenger stations in the terminal shall be determined through comparison in accordance with the convenience of passenger transportation, passenger traffic volume, passenger flow nature, exiting equipment condition, operation requirement, urban planning and local transportation condition, etc.

1 Generally, one passenger station may be established in the terminal which can serve as a confluence of different directions. At the intermediate station in cities which is of convenient transportation and can attract certain passenger flow, its passenger facilities and equipment may be reinforced based on actual requirement.

2 The railway terminal with large quantity of passenger transport capacity may be designed with two or more passenger stations. The passenger station shall be set or reserved quick and large-capacity urban public transport interfaces, to facilitate passengers to transfer to other means of transportation or to quick commute between the main passenger stations.

3 If the high-speed railway or intercity railway both be led in the terminal, the passenger station may be shared or constructed separately.

4.2.2 If there are two or more passenger stations in the terminal, their roles should be divided as follows:

1 To separately handle the originating/terminating trains that converges the lines, if conditions permit, to mutually handle the passenger trains passing through the passenger station.

2 If suburban passenger flow or intercity railway passenger flow is huge, the roles of stations may be divided as per long and short distance, intercity and suburban trains.

3 Based on traction type, the roles of stations shall be divided as per EMU trains and locomotive traction trains; if certain basis is available, the roles of stations shall be divided as per originating trains, terminating trains and passing trains, or as per originating trains, terminating trains and common passenger trains.

4.2.3 The selection for passenger station location shall comply with the following provisions:

1 To comply with the general view of railway terminal.

2 To stretch into or be close to passenger flow center, and to develop harmonically with the cities.

3 If several passenger stations are set in the terminal, they shall avoid being set at the same side of the city.

4 Integrated comparisons shall be made, and factors shall be taken into account in location determination such as urban planning, topographical and geological condition, building demolition, and land resource development.

4.2.4 For terminals located in large city or super-large city, if the railway line is of suburban passenger transportation demand, the equipment shall be provided for suburban passenger transportation. Stations without siding track may be established for passenger boarding & alighting in the terminal, if the station is close to large industrial park, residential area and main transfer stop of urban transportation.

II Marshalling Station

4.2.5 The number and configuration of the marshalling stations in the terminal shall be determined through comprehensive comparison in accordance with the traffic flow, traffic flow nature and directions, lead-in lines condition and role division of marshalling station in railway network, and local condition. For the station in the terminal with large number of loading/unloading operation, its equipment shall be reinforced as per the requirement of origin-destination transportation. The marshalling stations in the terminal should be intensively arranged. One marshalling station shall be set for the new terminal or the terminal primarily focusing on traffic flow transfer of the railway network. Under special conditions, two or more marshalling stations may be set in the terminal after technical and economic comparison and if one of the following conditions is satisfied:

1 Large number of transfer remarshalling traffic flows in railway network, and large number of local traffic flow intensively arriving and departing at industrial district and port district.

2 Lead-in lines converges at two places or more which are far from each other, there is certain number of reflected traffic flow and local traffic flow in the confluence, and scattered layout of remarshalling operation is preferable for traffic handling.

3 Large terminal range, many lead-in lines, scattered distribution of industrial enterprises, and large number of local rail traffic flow.

4.2.6 When two or more marshalling stations are set in the terminal, the quantity of work and work nature of every marshalling station shall be determined after economic and technical comparison according to the function division, traffic flow nature and locomotive routing in the railway network, and the following condition shall also be taken into account:

1 All transfer remarshalling operation shall be handled at one main marshalling station, and the marshalling for the reflected traffic flow of the lines connected with other marshalling stations in the terminal shall be handled within the marshalling station.

2 The function division of the marshalling station shall be performed by operation direction, which handles the remarshalling operation for traffic flow of different lines leading in the terminal. On particular occasion, it will undertake partial marshalling for traffic departure flow.

3 The function division of the marshalling station shall be performed by connecting lines, which handles the remarshalling operation for traffic flow of different lines approaching the terminal and outbound.

4 The integral labor division for marshalling station is generally as per connecting line or operation direction; most transfer remarshalling traffic flows shall be concentrated and handled in the main marshalling station.

4.2.7 The selection for marshalling station location shall comply with the following provisions:

1 Should be on the periphery of the urban planning area.

2 Shall be on the line where main traffic flow converges.

3 Short term development requirement shall be combined together with the long term development requirement, the operation demand of lead-in lines for design period shall be met, and potential development shall be taken into account in decision-making.

4 For the marshalling station mainly focusing on transfer remarshalling operation, its location shall guarantee the traffic flow on the main line to pass through the terminal in shortest possibly route. For the marshalling station which serves transfer and local marshalling operation, it location shall facilitate the operation of straight and reflected traffic flow for the transfer traffic flow, and the running distance shall be possibly reduced for the district transfer train in the service area. The marshalling station which serves the remarshalling of local traffic flow shall be close to the main industrial district or port area.

4.2.8 The passing freight trains should be handled in the marshalling station. If a large number of passing freight trains is not required to be handled in the marshalling, station may be built separately, whose location should be close to the marshalling station in order to share its locomotive facilities.

III Railway Logistics Center

4.2.9 The number, function division and configuration of the railway logistics center in the terminal shall be determined after comparison, and the principles of facilitating freight transportation and relative concentration shall be followed in accordance with freight traffic flow, freight natures, work nature, operation requirement, existing equipment condition, urban planning and local transportation conditions, etc. If the terminal is of large range and the cities involved are scattered, based on actual demand, several railway logistics centers of different feature and different grade may be set; the terminal located in middle or small cities may be equipped with the railway logistics center of small scale; at the residential area, industrial district around the terminal, or at the station near the satellite city, the logistics work station, acceptance station, acceptance point or trackless station may be set if necessary.

4.2.10 The railway logistics center shall meet the industry requirements, close to freight source, and be coordinated with urban industrial layout, industrial parks, logistics parks and transportation planning; the exiting freight yard located in urban area may be gradually transformed into the logistics center which directly serve the urban residents based on urban planning and adjustment, or the station yard may be comprehensively developed according to the property of the land nearby. In deciding the scale of railway logistics center, the service function, logistics development, and operation and development requirement shall be taken into consideration.

4.2.11 The railway logistic center should be designed as comprehensive logistics center; for the terminal located at big city, special-purpose railway logistics center may be set according to actual requirements.

4.2.12 The railway logistics center should be set at the loop line, the bypassing line or the connecting line, or may be set at the line led out from the marshalling station or the intermediate station, or may be set on the intermediate station if necessary. The selection for the location of railway logistics center shall comply with the following provisions:

1 To comply with the general view of railway terminal and urban logistics planning, to be correspondent with the productive forces layout of freight transportation, to adapt to the capacity of the connecting line or station, and to be close to technical operation station to ensure traffic flow smooth.

2 To be located at or close to industrial district, logistics parks, industrial and mining enterprises, port and pier, and other freight collecting and distributing centers, the market demand and multimodal transport shall be fully taken into consideration. According to "seamless" connecting requirement, multiple transportation means shall be reinforced to implement "door to door" transportation service.

3 The railway logistics center which serves the railway-river combined transportation shall be located at or close to the port.

4 The professional logistics center for hazardous freight shall be located at the suburb or at the downwind side of the urban prevailing wind direction, and shall be far away from the residential area or other environmental sensitive areas.

5 The railway logistics center with express delivery function area should be close to and easy to get access to the high-speed railway.

6 To be with good topographical, geological, hydrological and meteorological conditions, which makes it easy to connect with the external auxiliary facilities like urban roads, water, electricity and gas, etc.

IV Locomotive, Rolling Stock And EMU Equipment

4.2.13 Locomotive, rolling stock and EMU equipment shall be reasonably allocated according to the whole railway and regional planning, shall be possibly close to the marshalling station, the railway logistics center or the main passenger station according to the principle "centralized repair and distributed inspection", the construction scale shall be system matching and potential development shall be considered.

4.2.14 The locomotive equipment in the terminal shall be determined according to the passenger and freight locomotive routing for connecting lines, as well as the technical operation nature of the locomotives, the maintenance and servicing equipment for passenger and freight locomotives may be allocated according to the following requirements:

1 The maintenance equipment for passenger and freight locomotives in medium and small terminals shall be built at one site. For large terminal with complicated locomotive maintenance work, the maintenance equipment for passenger and freight locomotives may be built in different sites.

2 The marshalling station and the passenger station shall be equipped with locomotive servicing facilities, where many pairs of passenger trains are dragged by locomotives. If conditions permit and without many pairs of passenger trains, in between the passenger station and the marshalling station the locomotive servicing equipment may be stored and shared by passenger and freight trains, in addition, the special locomotive running track may be set.

3 The maintenance base for high-power locomotives shall meet the requirements for locomotive repairing in the section and related facilities layout.

4.2.15 The rolling stock equipment in the terminal shall be allocated according to the factors like the quantity of railway passenger and freight vehicles and the specified conditions for detaining rolling stocks. The freight car depot shall be built at the district where there is terminal's marshalling station, industrial station or harbour station with car breaking-up and marshalling operation, unloaded-car gathering operation and easy for car detaining operation. The passenger rolling stock depot shall be arranged in the passenger station with originating trains, terminating trains and with many allocated passenger trains, and should be built together with passenger train servicing depot.

4.2.16 For the passenger station with large number of passenger train originating and terminating operation, the longitudinal-type passenger car depot (shed) should be built nearby, which shall meet the requirements of the station layout plan and transportation development.

4.3 Station Approach Line Layout and Untwining

4.3.1 The station approach line layout shall comply with the following requirements:

1 The passenger train is connected to the passenger station through thelink line, the passenger train of the main direction shall not change running direction while passing though the terminal.

2 The freight train is connected to the marshalling station through the link line, the straight route shall be available for the main traffic flow to pass through the terminal.

3 The arrival track and departure track of the passenger train and freight train led in from different directions shall be separately connected to the passenger station and the marshalling station; the departure tracks may be combined appropriately and led out of the above stations according to the passing capacity of the sections, the work handling capacity of the station and the engineering conditions.

4 The link line which meets the operation requirements shall be available between the link lines, and between related stations in the terminals.

4.3.2 The station approach lines untwining may be designed as interchange untwining or plane untwining according to the traffic flow, traffic safety conditions, the requirement of trains running in different directions and in different categories, as well as the local conditions. The station approach lines untwining shall also be design based on the technical requirement of the plan and profile of the line, and in accordance with the urban planning, topographical and geological conditions. For the new terminal or the terminal with few link lines and most of the link lines are single-track confluence, the station approach lines may be designed as plane untwining.

4.3.3 The station approach lines untwining should be designed as per train operation directions (Figure 4.3.3-1). The station approach lines may be designed as per other untwining styles if the following conditions exist:

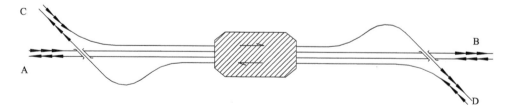

Figure 4.3.3-1 Sketch Map of Station Approach Line Untwining as per Train Operation Directions

1 For the passenger and freight shared station with little alternating quantity between lines, where single-track railway converges with double-track railway or two single-track railways converges, the station approach line may be designed as per line types (Figure 4.3.3-2), but the possibility shall be retained for reconstruction as per direction.

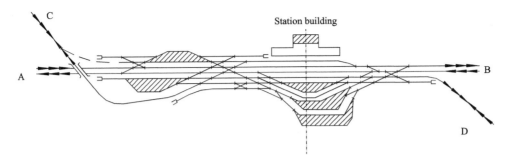

Figure 4.3.3-2 Sketch Map of Station Approach Line Untwining as per Line Types

2 If some sections or some station approach lines in the terminal is necessary to set special main line for certain trains, the untwining may be designed as per train types (Figure 4.3.3-3). When two or more above lines are designed as per train types for untwining, their special main line should be arranged as per directions, for some special main lines led in for single track and retain some plane crossing, the line may be arranged as per line.

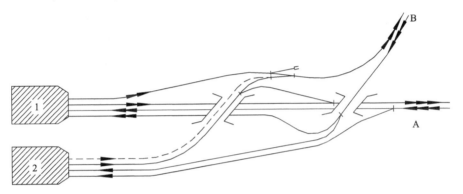

Figure 4.3.3-3 Sketch Map of Station Approach Line Untwining as per Train Types

4.3.4 The layout style of untwining line shall be determined through technical and economic comparison in accordance with running direction, train running condition, station layout and reduction of operation crossing in the station.

4.3.5 The station approach lines designed as per interchange untwining shall reserve location for new lead-in line, upgrade main line and connecting line.

4.3.6 The station approach lines designed as per plane untwining shall comply with the following requirements:

1 The access layout shall be flexible, and access crossing shall be distributed at the throat area at both ends.

2 Proper lines in the station shall also be served for train refuge.

3 The throat area layout shall be with appropriate parallel accessroute.

4 In front of the home signal the starting-stopping conditions shall be available.

4.4 Bypassing Line and Connecting Line

4.4.1 The following bypassing line or connecting line may be arranged or reserved in the general view of the terminal. Passenger trains may pass through the bypassing line and the connecting line if necessary:

1 The bypassing line at periphery of the terminal to bypass the city.

2 The bypassing line in the terminal to bypass some stations.

    3 The bypassing line in the terminal to enable the freight train to bypass the urban area.
    4 The connecting line to reduce the excessive running of reflected traffic flow.
    5 The bypassing line for national defense.

4.4.2 In the construction of the bypassing line at the terminal periphery, the requirement of traffic flow organization and locomotive routing for adjacent marshalling station shall be thoroughly studied, and the crossing and untwining for the bypassing line leading to track junction point shall be properly handled. The ruling gradient of the bypassing line and the effective length of the receiving-departure track for the stations shall match the standard of the connecting lines. The distribution of the dividing points of the bypassing line shall meet the required carrying capacity.

4.4.3 The locomotive equipment of the connecting lines should be shared for the bypassing line, if necessary, the locomotive equipment, train inspection and the locomotive crew shift changing equipment may be prepared at the track connecting station or the front station of the bypassing line.

4.4.4 The technical standard of the connecting line shall be determined according to the undertaken works, property, traffic flow, and topographical and geological conditions. For the connecting line for reflected trains passing through between the lead-in lines in the terminal, its length, plan and profile shall guarantee the train start-stop condition at the connecting line.

4.4.5 The connecting line in the terminal shall be designed to meet the requirement of train running over different lines.

5 Marshalling Station

5.1 General Requirements

5.1.1 The marshalling station can be divided into railway network marshalling station, regional marshalling station and local marshalling station. The marshalling station shall be designed according to the number of lead-in lines, work load and nature, engineering condition, urban planning and other requirements, through comprehensive comparison, the reasonable layout plan shall be selected and the potential development shall be reserved according to actual demand.

5.1.2 The marshalling station shall be overall planned and constructed by stages based on the increasing of traffic volume. The design of the present project shall facilitate operation, reduce construction cost and reduce the demolition engineering and operation disturbance for future expansion.

5.1.3 Under the premise that the passage capacity and remarshalling capability have been met, and the engineering investment cost and operation cost have been possibly reduced, the configuration of yard, shunting equipment and other devices of the marshalling station shall comply with the following requirements:

1 All components of the station shall cooperate with each other.

2 The station operation is of flexibility and flow process.

3 To reduce access road crossing and operation disturbance.

4 To reduce the running distance and duration of stay at the station for locomotives, rolling stocks and trains.

5 To be easy to adopt modern technical equipment.

5.2 Layout Plan of Marshalling Station

5.2.1 The marshalling station shall be designed as unidirectional layout plan or bidirectional layout plan through technical and economic comparison in accordance with remarshalling work load, reflected traffic flow, topographic condition, and station approach line layout. The hump direction of unidirectional marshalling station shall be determined in accordance with remarshalling traffic flow and its direction, topographic conditions and meteorological conditions. The capacity and layout style of the two systems for bidirectional marshalling station may be determined according to the actual demand.

5.2.2 The transverse-type layout plan of marshalling station with one receiving-departure yard and one shunting yard shared by both directions is applicable to the small marshalling station with small number of breaking-up and marshalling operations. If station building and topographical condition permit, the yard should be arranged at the side of the main line close to remarshalled traffic flow.

5.2.3 The transverse-type layout plan of marshalling station (Figure 5.2.3) with the bidirectional receiving-departure yards respectively paralleling at the both sides of the shared shunting yard may be applicable to the marshalling station with balanced bidirectional remarshalling traffic flow and with small number of breaking-up and marshalling operations, or may be applicable to the medium and small marshalling station with topographical difficulty and without large scale of development in the long term. If the tonnage rating of the connecting line is heavy, the plan and profile conditions of the connecting line shall be properly handled which is transferred from lead line of hump.

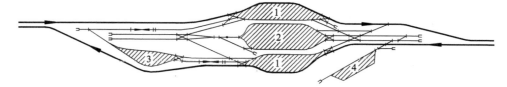

Figure 5.2.3 Transverse-type Layout Plan of Marshalling Station
1—Receiving-departure yard and transit yard; 2—Shunting yard; 3—Locomotive depot; 4—Rolling stock depot

5.2.4 The hybrid-type layout plan of unidirectional marshalling station (Figure 5.2.4), where bi-directionally shared receiving yard and shunting yard are arranged longitudinally and the departure yards are arranged at both sides of the bi-directionally shared shunting yard, may be applicable to the large and medium marshalling station with large number of breaking-up and marshalling operations or with topographical difficulty and large number of breaking-up and marshalling operations. If the remarshalling traffic flow is of bigger proportion to hump direction, some necessary measures shall be taken to balance the work load of the lead line at both sides of the shunting yard tail.

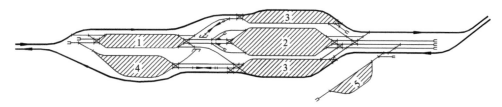

Figure 5.2.4 Hybrid-type Layout Plan of Unidirectional Marshalling Station
1—Receiving yard; 2—Shunting yard; 3—Departure and passing yard; 4—Locomotive depot; 5—Rolling stock depot

5.2.5 The longitudinal-type layout plan of unidirectional marshalling station (Figure 5.2.5), where bi-directionally shared receiving yard and shunting yard as well as departure yard are arranged longitudinally, may be applicable to the large scale marshalling station with large remarshalling traffic flow in hump direction and with large number of breaking-up and marshalling operations, and shall comply with the following requirements:

1 The line for receiving and departure of remarshalling trains reverse hump direction should be designed as vertical crossing.

2 The line for receiving and departure of remarshalling trains reverse hump direction should be designed as reverse receiving and reverse departure, and the conditions for receiving loop and departure loop shall be reserved. It may be designed as receiving loop and departure loop if the basis is available in short future.

3 If the unidirectional hybrid-type marshalling station is expanded into the unidirectional longitudinal-type marshalling station layout plan where the receiving yard, shunting yard and the departure yard are arranged longitudinally, the departure yard and passing yard for reverse hump direction may be retained.

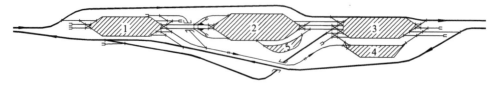

Figure 5.2.5 Longitudinal-type Layout Plan of Unidirectional Marshalling Station
1—Receiving yard; 2—Shunting yard; 3—Departure and passing yard; 4—Locomotive depot; 5—Rolling stock depot

5.2.6 For the unidirectional marshalling station with double humping operation, the receiving yard, shunting yard and the departure yard should be longitudinally arranged. According to the operation requirement of reflected traffic flow, some lines at the middle of shunting yard shall be designed as shared lines for sliding track at both sides. The tail part layout of the shunting yard and the configuration of marshalling facilities shall ensure the operation capability to be in line with the humping capability. The receiving track for remarshalling train in reverse hump direction should be designed as receiving loop. The departure route for remarshalling train in reverse hump direction should be designed as departure loop or reverse departure.

5.2.7 The hybrid-type layout plan of bidirectional marshalling station (Figure 5.2.7), where bidirectional receiving yards and shunting yards are arranged longitudinally, and the departure yards are arranged transversely at the outside of the shunting yard, may be applicable to the large scale marshalling station with large number of bidirectional breaking-up and marshalling operation, or with large number of bidirectional breaking-up and marshalling operation but with little reflected traffic flow and restricted by topographical conditions.

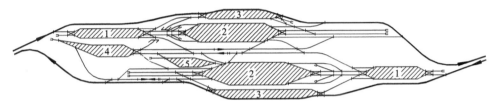

Figure 5.2.7 Hybrid-type Layout Plan of Bidirectional Marshalling Station
1—Receiving yard; 2—Shunting yard (marshalling-departure yard); 3—Departure and passing yard;
4—Locomotive depot; 5—Rolling stock depot

5.2.8 The longitudinal-type layout plan of bidirectional marshalling station (Figure 5.2.8), where bidirectional receiving yards, shunting yards and departure yards are arranged longitudinally, may be applicable to the large scale marshalling station with large number of bidirectional breaking-up and marshalling operation.

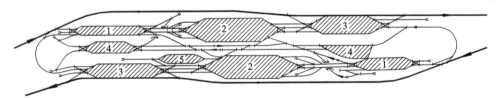

Figure 5.2.8 Longitudinal-type Layout Plan of Bidirectional Marshalling Station
1—Receiving yard; 2—Shunting yard; 3—Departure and passing yard; 4—Locomotive depot; 5—Rolling stock depot

5.2.9 When longitudinal-type layout plan of the receiving yard and shunting yard are used and the remarshalling traffic flow in hump direction is big but with simple group number or mainly with local traffic flow, together with few departure direction for connecting line, the departure yard may not be arranged in hump direction according to actual conditions, all departure operation of rolling stock may be handled at the marshalling-departure yard. When the receiving-departure yard or the departure yard is transversely arranged with the shunting yard, some marshalling-departure tracks may be designed at the shunting yard.

5.3 Main Facilities and Equipment

5.3.1 The main line position in the marshalling station shared by passenger trains and freight trains shall be designed as outer enclosure type or one-side type according to the section design speed, traffic volume, location of passenger station, layout of railway logistics center and branch line, as well as the adopted layout plan. On the main line within the range of marshalling station, the passenger stop point may be arranged for passenger trains and commuter trains according to actual demand. If the commuter trains need to stop near the yard or depot within the marshalling station, the platform may be set at appropriate position.

5.3.2 The position of the passing yard shall be determined according to train operation smoothness, easy to pick and drop operation, convenient locomotive entering/exiting depot, little disturbance to marshalling operation, saving of equipment and staff capacity, and shall comply with the following requirements:

   1 The passing yard of transverse-type marshalling station should be set next to the receiving-departure yard; and the passing yard of hybrid-type and longitudinal-type marshalling station should be set next the departure yard.

   2 The mutually accessible route shall be set for the passing yard with its adjacent receiving yard, departure yard or the receiving-departure yard.

   3 The passing yard may not be built separately if there is few passing trains, the train operation shall be handled at corresponding rolling stock yard.

5.3.3 The tail part of shunting yard of the marshalling station shall adopt the central control of shunting access road. If there is large number of marshalling operation for multiple group trains, pick-up and drop-off trains and district transfer trains, the small capacity hump or auxiliary shunting yard shall be set at the tail part of the shunting yard according to the requirement of marshalling capacity and actual conditions.

5.3.4 The marshalling station shall, based on actual condition, connect some lines of the shunting yard with the main line.

5.3.5 The marshalling-departure track should be intensively set at the lines outside of the shunting yard, its yard runout throat should be appropriately added with parallel access road, and necessary safety protection facilities shall be established according to specific condition.

5.3.6 For the marshalling station with longitudinal-type layout for shunting yard and departure yard, the distance from the harness tail turnout of half lines in the shunting yard to the outmost turnout at the yard entering side of the departure yard should be half of the effective length of the receiving-departure track, and may be shortened under difficult conditions.

5.3.7 In order to ensure the reflected traffic flow of the bidirectional marshalling station to transfer from the shunting station of one set of system to the hump front yard of the other system, the connecting line should be set between the two systems; if the reflected traffic flow is big, the backup line or exchange yard shall be set between the two systems. On the basis of reflected traffic flow, nature of marshalling station, specific conditions, layout of station approach line of marshalling station, the corresponding system should be capable of reverse train receiving or departure in main reflected traffic flow direction.

5.3.8 The configuration of locomotive equipment for marshalling station shall be determined through technical and economic comparison according to marshalling station layout plan, locomotive operation condition and local conditions, and shall comply with the following requirements:

   1 The transverse-type marshalling station with the bidirectional receiving-departure yard paralleling at the both sides of the public shunting yard should be set at the hump end; the locomotive depot of the unidirectional hybrid-type marshalling station should be set at the one side of reverse hump direction near the receiving yard; the locomotive depot of the unidirectional longitudinal-type marshalling station should be set at departure yard within intensive arriving and departure operation or at the one side of reverse hump direction near the receiving yard, and shall be set at the one side of reverse hump direction near the shunting yard if receiving and departure loop is adopted.

2 The locomotive depot of bidirectional marshalling station should be set between the two systems and shall be close to the one end of the departure yard and the passing yard with large number of traffic flow, and at the other end the locomotive servicing equipment may be set if necessary.

3 If a large number of passing trains go through the passing yard without technical operation, the necessary locomotive servicing equipment and corresponding facilities may be set near the passing yard after technical and economic comparison.

5.3.9 The rolling stock depot of unidirectional marshalling station should be set near the tail part of the marshalling station; and the rolling stock depot of bidirectional marshalling station shall be set between the two systems and should be close to the tail part of the main unloaded train direction. The freight car repair point at station should be set near the tail part of the shunting yard. The rolling stock depot and freight car repair point at station should be combined into one if conditions permit.

5.3.10 If the refrigerator car needs to be fueled up in the marshalling station, the filling-up point shall be set at appropriate place, and automobile way shall be constructed outside the line where the filling-up point is located.

5.3.11 The packaging equipment and transshipment facilities for freight should be arranged in the railway logistics center near the marshalling station, if the operation work load is big and there is no railway logistics center nearby, they may be arranged at the side of the shunting yard where there are rolling stock maintenance facilities.

5.3.12 The railway logistics center and the branch line with large number of loading/unloading operation shall not be directly connected with the marshalling station with large number of breaking-up and marshalling operations, if they have to be connected, another yard or station should be arranged for connection at proper place where it is easy for car taking-out and placing-in operations and crossings in marshalling station should be minimized.

5.3.13 The transmission equipment for various documents shall be equipped for all yards of the marshalling station according to work requirement.

5.4 Number and Effective Length of Station Tracks

5.4.1 The number of tracks of the receiving yard, receiving-departure yard and the departure yard in the marshalling station shall be analyzed, calculated, simulated and decided according to number of trains, type of trains, frequency of arriving and departing trains, as well as the station technical operation process, the Table 5.4.1 shall be referred to in designing.

Table 5.4.1 **Number of Tracks of Receiving Yard, Receiving-Departure Yard and Departure Yard in Marshalling Station**

Number of receiving-departure trains	Number of tracks
≤18	3
19~30	3~4
31~42	4~5
43~54	5~6
55~66	6~7

Table 5.4.1 (continued)

Number of receiving-departure trains	Number of tracks
67~78	7~8
79~90	8~9
91~102	9~10

Notes: 1 The number of receiving-departure trains listed in the table refers to the sum of arriving trains and departure trains in different directions of the rolling stock yard.
     2 The range of number of lines in the table may be valued in accordance with the correspondent value of the train number, if it is out of the range of the table, it may be calculated as per the extrapolation.
     3 For the receiving yard, departure yard and receiving-departure yard with certain number of district transfer trains, the number of lines may be appropriately reduced based on the values listed in the above table.
     4 For the passing yard with handling for swap trailer or without swap trailer, the number of lines may be valued as per the lower limit or upper limit of the values listed in above table (the arriving and departure train shall be calculated as 1 set).
     5 The locomotive running track may be set based on demand.
     6 If the above table is applied, and there are 3 or more connecting lines at the rolling stock yard, 1 more line may be added. For front hump receiving yard, at least 2 lines shall be considered for every connecting direction, if few number of trains to be handled, the total line of the receiving yard may be suitable reduced.

5.4.2 The number and effective length of the line in the shunting yard of the marshalling station shall be determined in accordance with the line purpose, formation number of train formation plan, the traffic flow of every day and night for one formation number and the effective length of the receiving-departure track. The number and effective length of the lines for the shunting yard may be selected from Table 5.4.2.

Table 5.4.2 Number and Effective Length of Tracks of Shunting Yard

No.	Function of track	Number of tracks	Effective length
1	To assemble marshalling non-stop train, transit train or section trains	One track for one formation number based on marshalling plan; one more track may be added if the traffic flow of one day and night exceeds 200; one track may be shared by two formation number if traffic flow is small	As per the effective length of the receiving-departure track, some tracks may be slightly smaller than the above length
2	To assemble unloaded trains	Shall be determined as per type of unloaded train and one day-and-night traffic flow and with reference to the provision 1, but at least one track shall be prepared for every station with unloaded trains assembled (or every shunting yard of bidirectional marshalling station)	As per the effective length of the receiving-departure track, some tracks may be slightly smaller than the above length
3	To assemble the marshalling-departure track of marshalling non-stop train, transit train or section trains	Heavy-loaded trains shall be determined as per formation number and traffic flow of one day and night, and unloaded trains shall be determined as per train type and traffic flow of one day and night. 2 more tracks shall be added if traffic flow is 150~350, and 1 more track may be added if traffic flow is over 350	As per the effective length of the receiving-departure track
4	To assemble marshalling pick-up and drop-off trains	1 track shall be set for every connecting direction, and may be appropriately added for passage of major pick-up and drop-off trains according to traffic flow	1 track in every connecting direction, the length shall be the length of the train set plus 80~100 m
5	To assemble local transfer trains (including the marshalling-departure track of local transfer trains)	Shall be determined respectively according to the formation number of the formation plan and the traffic flow of one day and night: 1 track shall be set for 250 and below; 2 tracks shall be set for over 250	As per the length of the train set plus 80~100 m

Table 5.4.2(continued)

No.	Function of track	Number of tracks	Effective length
6	Exchange trains (the trains requiring repeated breaking-up)	At least 1 track for every shunting yard of bidirectional marshalling station, the unidirectional marshalling station with double humping may be determined based on layout plan	To be determined based on traffic flow
7	Operation rolling stock of the station	To be determined according to unloading place and number of rolling stock unloading	To be determined based on traffic flow
8	Packaged or transshipped trains	1	May be smaller than the effective length of the receiving-departure track
9	Rolling stock to be repaired	1	May be smaller than the effective length of the receiving-departure track
10	Rolling stock with out-of-gauge freight or not allowed to pass through the hump	1	May be smaller than the effective length of the receiving-departure track
11	Rolling stock loaded with hazardous freight	1	May be smaller than the effective length of the receiving-departure track

Notes: 1 The effective length of the track in the serial number 1,2 may be determined according to the requirement of assembling marshalling single formation trains.
    2 The calculating point for the effective length of the shunting track: the first retarder location terminal of the entrance of the shunting track (or the insulation joint behind) to the fouling post at the tail part of the shunting track (or the starting signal of the shunting track).
    3 If the effective length of the receiving-departure track is 1 050 m, 50 more cars shall be added to the traffic volume in Column "Number of tracks" numbered 1,3,5.
    4 The tracks numbered 8~11 may be separately arranged or combined according to needs.

5.4.3 The shunting neck in the marshalling station used for train breaking-up and marshalling operation shall be determined in accordance with the functions of shunting area, workload and working method. According to requirements, shunting neck may be arranged at passing yard for trains drop-and-pull operation or weight changing. The effective length of the shunting neck used for train breaking-up and marshalling operation may be designed based on the effective length of the receiving-departure track plus 30 m. If the topographical condition is difficult and the work quantity is little, the effective length of the shunting neck mainly applied for marshalling may be determined according to the working methods applied, but shall not be less than the 2/3 of the effective length of the receiving-departure track.

6 District Station

6.1 Layout Plan of District Station

6.1.1 The district station shall adopt transverse-type layout plan (Figure 6.1.1-1 and Figure 6.1.1-2). If sufficient basis is available, the longitudinal-type layout plan for passenger and freight service may be adopted (Figure 6.1.1-3) or the layout plan of one level with three yard.

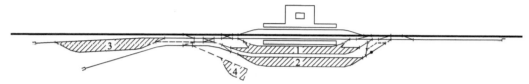

Figure 6.1.1-1 Transverse-type Layout Plan of District Station for Single-track Railway

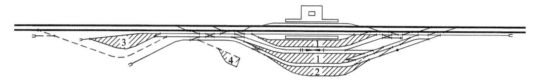

Figure 6.1.1-2 Transverse-type Layout Plan of District Station for Double-track Railway

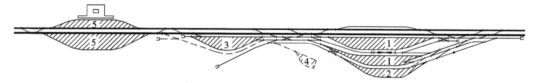

Figure 6.1.1-3 Longitudinal-type Layout Plan of Passenger and Freight District Station for Double-track Railway
1—Receiving-departure yard; 2—Shunting yard; 3—Locomotive depot; 4—Logistics center; 5—Passenger rolling stock yard

6.1.2 The layout plan of district station shall be determined in accordance with the number of incoming lines, traffic volume, transportation nature, station operation feature, passenger and freight locomotive routing, urban planning, topographical condition, geological condition, transportation requirement and technical and economic reasonability, and the following requirements shall be complied with:

    1 The district station for single-track railway shall adopt transverse-type layout plan. Other reasonable layout plans may also be adopted if there are many incoming lines from different directions and the traffic volume is big.

    2 The district station for double-track railway should adopt transverse-type layout plan. If there are incoming lines with large traffic volume, many passenger trains serving as the origin and destination point of passenger and freight locomotive routing, and the topographical condition is suitable as well, the layout plan of one level with three yard and other suitable layout plans may also be adopted or reserved.

    3 If large engineering works may be brought about owing to the above mentioned layout plans are adopted for the reconstructed district stations, or the local conditions are not suitable, the existing layout plan may be retained or other suitable layout plans may be adopted after technical and economic comparison.

6.2 Main Facilities and Equipment

6.2.1 The passenger station building shall be at the side of the main residential area of the city, and the location of the station building shall be in line with the urban planning. The intermediate platform shall not be near to the main line. In between the intermediate platform of single-track railway and the basic platform there should be 3 tracks, and 4 tracks for double-track railway; for the transverse-type district station of double-track railway only for shift changing of locomotive crews, one receiving-departure track may be added at the same side of the station building.

6.2.2 The receiving-departure track for passenger trains shall be capable of handing the receiving and departure of common freight trains, and the receiving-departure track should be designed as two-way route.

6.2.3 The throat area of the district station shall ensure the station with necessary passing capability, remarshalling capability, operation safety and operation efficiency, and shall comply with the following provisions:

1 The layout of the throat area shall meet the necessary parallel operation route, and the parallel route shall not be less than the number of main parallel route specified in Table 6.2.3.

2 Some lines of the shunting yard shall be connected with the main line. At the station with large number of remarshalling operation, some lines of the receiving-departure yard shall be capable of handling the parallel operation of the shunting and train receiving and departure.

3 The throat area shall be compactly designed, the number of turnouts on the conflicting route and the main line should be reduced, and the shunting stroke shall be the shortest.

4 Where the shunting route is connected with the main line, an isolated route shall be set.

Table 6.2.3 Parallel Operation Quantity at Throat Area

Layout plan	Condition		Location of throat area	Parallel operation quantity	Content of parallel operation
Transverse-type	Single-track railway	Number of parallel running trains is 18 pairs or less	Non-locomotive depot side	2	Train arriving (departure), shunting
			Locomotive depot side	2	Train arriving (departure), locomotive exiting (entering) section
		Number of parallel running trains is more than 18 pairs	Non-locomotive depot side and locomotive depot side	3	Train arriving (departure), locomotive exiting (entering) section, shunting
	Double-track railway		Non-locomotive depot side	3	Train arriving, train departure, shunting or train arriving (departure), locomotive exiting (entering) section, shunting
			Locomotive depot side	4	Train arriving, train departure, locomotive exiting (entering) section, shunting or train arriving (departure), locomotive exiting section, locomotive entering section, and shunting
Longitudinal-type	Double-track railway		Middle part	4	Down train departure (passing), up train departure, locomotive exiting (entering) section, and shunting

6.2.4 The railway logistics center of the district station should be set at the side of the shunting yard.

6.2.5 The location of the new locomotive depot for the district station shall be determined in accordance with operation requirement, scale, and topographical, geological, hydrological and drainage conditions. The locomotive depot should be set at the right end opposite to the passenger station building if the transverse-type layout plan is used for the station. For the district station with cycling operation or long locomotive routing and for shift changing of locomotive crews, the locomotive servicing facilities may be set for the receiving-departure track of the receiving-departure yard according to actual needs.

6.2.6 The rolling stock depot and freight car repair point at station for the district station should be set outside of the shunting yard or other suitable places.

6.2.7 The connection of branch lines for the district station shall be overall planned. If there are several branch lines requiring to be connected, they should be combined and introduced. The branch line may be connected at the shunting neck of the logistics center, the secondary shunting neck of the shunting yard, the shunting yard or other station tracks, and the branch line should be connected to the receiving-departure yard if the traffic volume is big and the whole train arriving and departure need to be handled.

6.2.8 The stabling siding shall be provided for the district station where passenger train set stays, and the train set shall be easily transferred to the receiving-departure track.

6.3 Number and Effective Length of Station Tracks

6.3.1 The number of the receiving-departure track for the district station shall be determined in accordance with the type, nature, num and operation style of the rolling stock, and the Table 6.3.1 may be referred to in the design process.

Table 6.3.1 Number of Receiving-Departure Tracks for District Station

Converted pair number of trains	Number of receiving-departure tracks (except main line and locomotive running track)
≤12	3
13~18	4
19~24	5
25~36	6
37~48	6~8
49~72	8~10
73~96	10~12
>96	12~14

Notes: 1　The number of receiving-departure track in the above table may be valued as per converted pair.
   2　For the district station with incoming lines from two or more directions (including the industrial siding as per operation handling), considering the synchronous arriving and departure of trains, the number of receiving-departure track may be appropriately added.
   3　If the tracing train working diagram is applied, one receiving-departure track shall be added.
   4　The dead-end type main line of the district station shall be calculated as the receiving-departure track.
   5　The number of receiving-departure track for freight trains at passenger and freight longitudinal district station shall refer to the above table after deduction of the converted pair of passenger trains, and the number of receiving-departure track for passenger trains shall be valued as per the Table 9.1.9 of this Code. The number of receiving-departure track for "one level with three yards" district station shall refer to the above table based on the converter train pair of up and down branch yard.
   6　The converted train pair in one direction of the district station is the sum of the pair of passenger and freight trains in all directions (can be calculated by the number of trains arriving and departure in this direction divided by 2) multiplying by the corresponding converted coefficient. When the above table is referred to for deciding the number of receiving-departure track, the dead-end type district station shall be determined by the total converted pair of all directions at the receiving-

departure end, but it can be appropriately reduced; the passing type district station shall be determined by the total converted pair of all directions divided by 2. The conversion coefficient of train pair: non-stay train, transit train, and local transfer train is 1, the non-stay train, transit train, section train, pick-up and drop-off train and fast freight train is 2, originating train and destination trains is 1, the local transfer train immediately turning around is 0.7, the passenger train stopping at station is 0.5, the freight train with shift change of locomotive crew and without train inspection is 0.3, and the freight train and passenger train without stopping in station are excluded.

6.3.2 One locomotive running track shall be set in the district station where locomotives passes through the rolling stock yard for 36 times or above every day and night.

6.3.3 In transverse-type district station, locomotive waiting track shall be arranged at the throat area not connected to the locomotive depot; in longitudinal type district station, locomotive waiting track shall be arranged at the throat area at the departing end of the receiving-departure yard on the opposite side of the locomotive depot. The locomotive waiting track may not be set for the transverse-type district station of single track railways with fewer locomotives to be changed or with difficulty for reconstruction. The locomotive waiting track should be dead-end type, and may be through type if necessary. The effective length of the locomotive waiting track: for the dead-end type, it shall be 45 m, under difficult condition, it shall not be shorter than the locomotive length plus 10 m; for the through type, it shall be 55 m, under difficult condition, it shall not be shorter than the locomotive length plus 20 m. For double-locomotive traction, the above mentioned effective length shall be added with one locomotive length.

6.3.4 The quantity and effective length of the shunting track for district station shall be determined according to the direction, quantity of connecting railway, number of operation trains, shunting operation method and train marshalling plan, and the following provisions shall be complied with:

1 No less than one track in every connecting direction and it may be appropriately added in directions with more traffic flow. The effective length shall not be shorter than the effective length of the receiving-departure track.

2 The waiting track for operation rolling stock shall not be less than one track; the waiting track for rolling stocks to be repaired shall be one track, and one track may be shared if there is only with a small number of rolling stocks; one more track may be added if there is siding connection and with much traffic flow; one waiting track shall be set for dangerous freight rolling stock if such rolling stocks are to be parked. The effective length of above mentioned shunting track shall be determined according to the max number of the rolling stocks assembled on the track.

6.3.5 One shunting neck shall be set at both ends of the shunting yard of the district station. When the quantity of breaking-up and marshalling operation is less than 7 trains for every day and night, the secondary shunting neck may be set later. The effective length of the main shunting neck shall not be less than the effective length of the receiving-departure track, and it may be appropriately reduced if there is only axle load adjustment on the shunting neck. The effective length of the secondary shunting neck should not be less than the effective length of the receiving-departure track, and it may be half of the effective length of the receiving-departure track with less number of shunting operation. If the track with less traffic capacity or siding is connecting at the station, and the plan and profile is suitable for shunting operation, it may be used as the secondary shunting neck.

6.3.6 One track shall be set for locomotive entry and exit each between the locomotive depot and the arriving departure yard of the transverse-type district station, when the number of locomotive entry and exit for every day and night is less than 60 times, one track shall be set. If other layout plans are adopted, the number of locomotive entry and exit depot shall be determined according to actual situation.

7 Intermediate Station and Combination & Disassembly Station

7.1 Intermediate Station

7.1.1 The transverse-type layout plan shall be adopted for intermediate station. The layout plan of the intermediate station for high-speed railway shall be arranged as per Figure 7.1.1-1, the layout plan of the intermediate station for intercity railway shall be arranged as per Figure 7.1.1-2, and the layout plan of the intermediate station for mixed traffic railway shall be arranged as per Figure 7.1.1-3 or Figure 7.1.1-4. Other suitable layout plans may be used under difficult conditions.

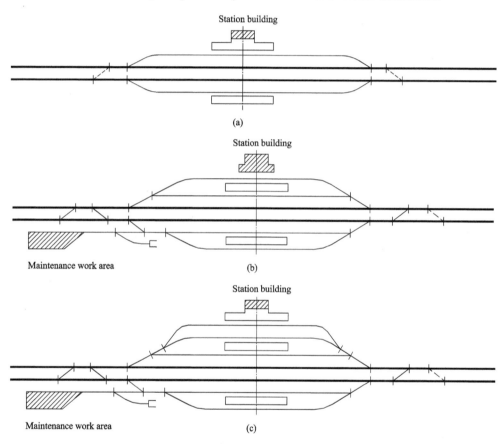

Figure 7.1.1-1 Layout Plan of Intermediate Station for High-speed Railway

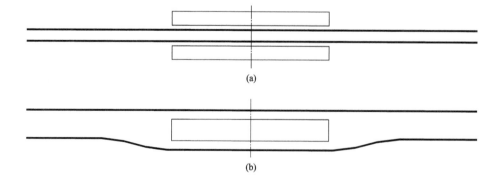

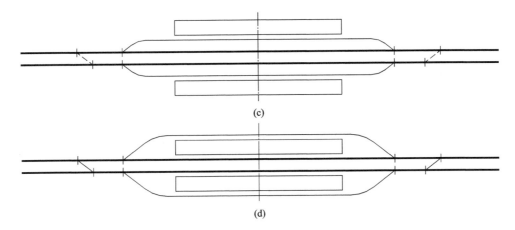

Figure 7.1.1-2　Layout Plan of Intermediate Station for Intercity Railway

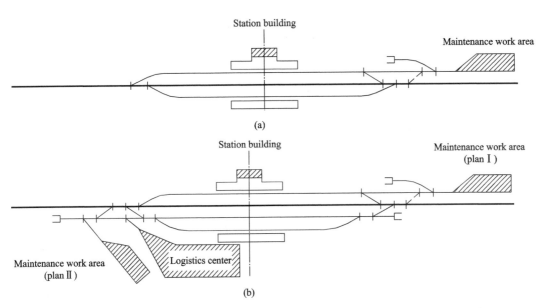

Figure 7.1.1-3　Layout Plan of Intermediate Station for Single-track Mixed Traffic Railway

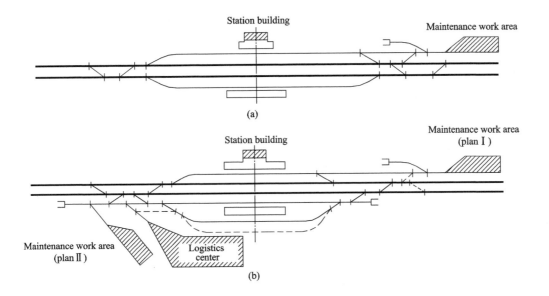

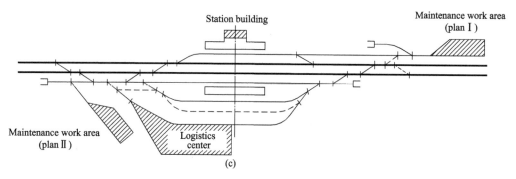

Figure 7.1.1-4 Layout Plan of Intermediate Station for Double-track Mixed Traffic Railway

7.1.2 The number of the receiving-departure track for the intermediate station shall comply with the following provisions:

1 2~5 receiving-departure tracks may be set for the intermediate station of high-speed railway.

2 For the successive 3~5 intermediate stations of the intercity railway or within every 20~30 km, one station shall be set with receiving-departure tracks and two receiving-departure tracks may be considered. Other intermediate stations may not be set with receiving-departure tracks.

3 Two receiving-departure tracks shall be set for the intermediate station of mixed traffic railway, and three receiving-departure tracks may be set in case of more operation workload. More receiving-departure tracks shall be added for the following intermediate stations based on the actual need:

1) Station of administration boundary, front station of marshalling station, helper locomotive origin/destination station, rolling stock technical inspection station for long and steep downgrades and locomotive crew transfer station.
2) The intermediate station with lead-in lines over two directions or connected with sidings.
3) The intermediate station with servicing and marshalling operation for pick-up and drop-off trains.
4) The intermediate station for operation of locomotive turning round or passenger train immediate turning back.

7.1.3 The arrangement of crossover between the two main lines of the intermediate station throat shall comply with the following provisions:

1 For the intermediate station of high-speed railway with originating/terminating train operation or EMU inspection train turn-back operation, the departure operation throat area shall be arranged with two sets of single crossover to form a wing crossover, while the non-departure operation throat area should be arranged with one set of single crossover. For other stations, one set of single crossover should be set at each throat area to form a wing crossover, and the orientation of the wing crossover should be determined based on the location of the maintenance work area. Crossover may not be set for the station where there is no maintenance work area and it is closer to the adjacent stations at both directions.

2 For the intermediate station of intercity railway with originating/terminating train operation, the departure operation throat area shall be arranged with two sets of single crossover to form a wing crossover, while the non-departure operation throat area should be arranged with one set of single crossover. For the successive 3~5 stations or within every 20~30 km, one station shall be set with auxiliary track and crossover.

3 Two sets of crossover should be set between the two main lines at both throat ends of the double-track intermediate station for mixed traffic railway and heavy-haul railway, namely, one set of crossover shall be set respectively at each end, the other 2 sets of crossover may be set or reserved based on the operational need.

7.1.4 The logistics center of the intermediate station shall be determined in accordance with main source of goods, freight flow direction, environmental protection, urban planning, topographical and geological conditions, and shall comply with the following provisions:

1 The logistics center should be established at the opposite side of the station building; it may be located at the same side of the station building if topographical condition is difficult and with less logistics flow.

2 The freight track shall be set if the annual workload of arrival and departure operation is huge and mass freight has to be handled, the loading/unloading conditions shall be met for the whole train, and the main freight flow direction shall be of direct passing through transportation condition.

3 The logistics center shall be connected with easy, safe and convenient access roads.

7.1.5 The setting of the shunting neck for the intermediate station shall comply with the following provisions:

1 The logistics center with goods loading/unloading track shall be set with shunting neck.

2 If the intermediate station is connected with siding, and the shunting operation conditions are met, the siding may be used for shunting operation.

3 If the siding is used for shunting operation, the home signal shall be shifted outwards and the distance shifted should not be more than 400 m. The horizontal and vertical profile as well as the observation condition shall comply with the requirement of shunting operation. The curve radius shall not be less than 300 m and the gradient shall not be more than 6‰ under difficult conditions.

4 The effective length of the shunting neck should not be smaller than half of the freight train which runs in the district; under difficult condition or with less operation in the station, it shall not be shorter than 200 m.

7.1.6 The intermediate station with locomotive turning round operation and locomotive allocation operation shall be set with locomotive waiting line and necessary servicing equipment.

7.2 Combination and Disassembly Station

7.2.1 The transverse-type layout plan shall be adopted for combination and disassembly station, as per Figure 7.2.1-1 or Figure 7.2.1-2.

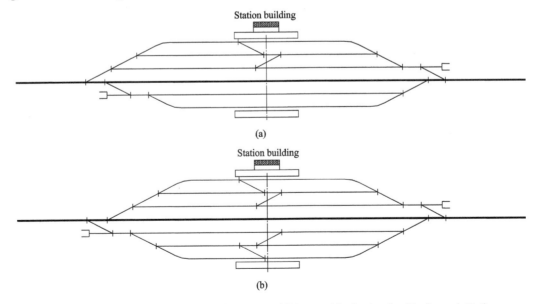

Figure 7.2.1-1 Layout Plan of Combination and Disassembly Station for Single-track Railway

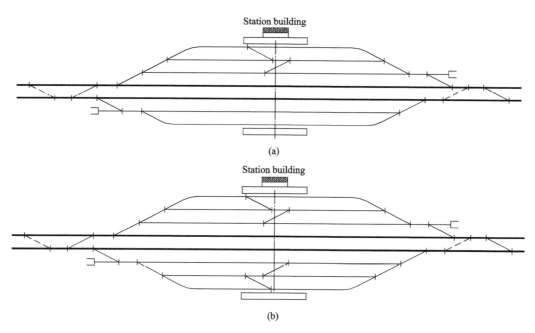

Figure 7.2.1-2　Layout Plan of Combination and Disassembly Station for Double-track Railway

7.2.2　The receiving-departure track of the combination and disassembly station should be regularly set according to the running directions of loaded train and unloaded train.

7.2.3　The number of the receiving-departure track for the combination and disassembly shall be determined according to the operation times of departure, coupling, disassembling, locomotive running time and train inspection, with the general consideration of each combination and disassembly track handing of 6～8 trains.

7.2.4　The length of both sections split by middle switches for combination and disassembly, and/or the length between the throat area and the middle switch shall satisfy the effective length of the train which requires to be combined or disassembled.

7.2.5　At the departing throat area of the receiving-departure track for the combination and disassembly where locomotives are intensively assembled, the locomotive waiting track shall be set, and the effective length of the locomotive waiting track shall be determined based on the maximum number of locomotives required to be parked.

7.2.6　The layout of the station track for the combination and disassembly station should adopt the layout style with one locomotive running track located between two combination and disassembly receiving-departure tracks, and the middle switches shall be set between the receiving-departure track and the locomotive running track based on the length of the combination and disassembly trains.

8 Passing Station and Overtaking Station

8.1 Passing Station

8.1.1 The transverse-type layout plan shall be adopted for the passing station, as per Figure 8.1.1. Other suitable layout plans may be adopted under difficult conditions.

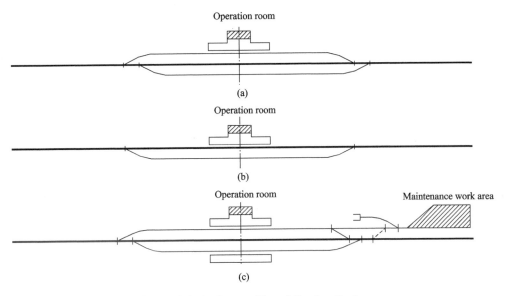

Figure 8.1.1 Layout Plan of Passing Station

8.1.2 Two receiving-departure tracks shall be set for the passing station. One receiving-departure track may be set in case of less traffic or under difficult conditions, but it shall not be set successively.

8.1.3 When one receiving-departure track is set for the passing station, it should be set at the opposite side of the operation room.

8.2 Overtaking Station

8.2.1 The transverse-type layout plan shall be adopted for the overtaking station, as per Figure 8.2.1. Other suitable layout plans may be adopted under difficult conditions.

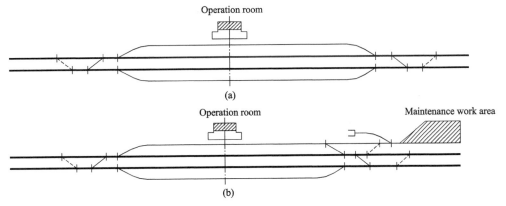

Figure 8.2.1 Layout Plan of Overtaking Station

8.2.2 Two receiving-departure tracks shall be set for the overtaking station.

8.2.3 Crossover may not be set for the overtaking station of high-speed railway if there is no maintenance work area in the station and it is closer to the adjacent stations. The crossover shall be set under other circumstances. One set of crossover shall be arranged respectively between the two main lines at each end of the throat for the overtaking station of mixed traffic railway, forming "big wing type" crossover. An additional set of crossover should be reserved respectively at each end of the station throat if conditions are available. The overtaking station with "big wing type" crossover shall not be set successively.

9 Passenger Station, Passenger Facilities and Equipment and Passenger Car Depot (Shed)

9.1 General Requirements

9.1.1 The establishment of passenger station shall be determined in accordance with the features of the city such as scale, status, nature, passenger traffic volume, conditions of existing passenger facilities and equipment, urban planning and the connection of urban traffic system.

9.1.2 The passenger station shall be divided into four categories according to maximum number of accumulated passengers or passengers departed at peak hours, i. e. super-large passenger station, large passenger station, medium passenger station and small passenger station. The scale of the facility of the passenger station shall be determined based on the category of the passenger station.

9.1.3 The design of passenger station shall comply with the following provisions:

1 To make overall consideration for the urban development and integrated transportation, to construct the easy transfer and smooth connected comprehensive transportation terminal centered by the passenger station, and to realize the "zero distance" transfer requirement.

2 Separate operation of passenger and freight transport should be adopted for the main passenger station located in large city, namely, the main line for freight trains shall bypass the passenger station or the freight train connecting line shall be set to shunt the freight trains.

3 The plane layout of the passenger station shall meet the requirement of quick and smooth running of passenger trains, the capacity of receiving-departure tracks and throat capacity of the station shall be in line with the sectional through capacity of the line, and the parallel operation requirement shall be met if there are multiple technical operations.

4 The general layout of the station shall be systematically designed according to the station streamline and the functional requirement of operation facilities. Both the function and capacity shall be matched, and considerations shall be taken to the passenger station that serves as the land integrated development platform.

5 The station vertical profile and elevation design shall be determined through economic and technical comparison in accordance with topographical and geological condition and comprehensive transportation planning of station region.

9.1.4 If the passenger station is shared by high-speed railway, intercity railway and mixed traffic railway, the separate lines and separate yards layout should be adopted. For the passenger station of high-speed railway or intercity railway lead-in with multiple directions, the car yard layout shall be reasonably determined according to the functional orientation of the lead-in lines, and separate yards should be arranged. A combined yard layout may also be adopted if there is less immediate turning round passenger trains or only with a third lead-in direction.

9.1.5 When there are several passenger stations in the terminal or local region, the passenger stations shall be of suitable division of work, and interconnected.

9.1.6 For the passenger station with passing trains, the setting of the main line shall comply with the following provisions:

1 For the passenger station of double-track railway, when the passenger car servicing depot and the passenger station are longitudinally arranged between the two main lines, the two main lines should be set outermost at the opposite side of the station building, and between the first platform and the second platform, respectively. When the passenger car servicing depot and the passenger station are longitudinally arranged at the same side of the station building, one of the main lines should cross the station, and the other one should be set outermost at the opposite side of the station building.

2 For the passenger station of double-track railway, when the section design speed is 120 km/h or above, the layout plan of two main lines parallel crossing should be adopted.

3 For the passenger station of single-track railway, the main line passing through freight trains should be set outermost at the opposite side of the station building.

9.1.7 The layout of station throat area shall satisfy the requirement of train receiving and departure, and the quantity of turnout on the main line shall be reduced. When there are several lead-in lines or running track of depot(shed) are connected, the throat area layout at the lead-in end shall meet the requirements for parallel operation of train arrival(departure), taking-out and placing-in of passenger train-set, EMU exiting and entering depot(shed), and locomotive exiting and entering depot.

9.1.8 Passenger locomotive depot, passenger car depot(shed) and the relevant running track and locomotive waiting track shall be set or reserved for the passenger station, and grade separation should be set for crossings between the lines of exiting/entering depot(shed) and the main line.

9.1.9 The number of receiving-departure tracks for passenger trains shall be determined in accordance with the factors like the pairs of passenger trains and their nature, number of lead-in lines and the technical operation process of the station.

1 The number of receiving-departure tracks for stations where most passenger trains are pulled by locomotive may be selected from Table 9.1.9.

Table 9.1.9 Number of Receiving-Departure Tracks for Stations Where
Most Passenger Trains are Pulled by Locomotive

Converted pair number of originating/terminating passenger trains	Number of receiving-departure tracks(line)
≤12	3
13~24	3~5
25~36	5~7
37~50	7~9

Notes: 1 The number range of receiving-departure tracks may be valued based on the pairs of the trains.

2 For the number of receiving-departure tracks for passenger station with passenger trains stopover, the stopover trains may be converted into originating/terminating trains and valued from the above table, and each pair of ordinary speed stopover train may be converted into and calculated by 0.5 pair of originating/terminating train. When the receiving-departure track is shared by ordinary speed passenger trains and EMUs, EMUs may be converted into ordinary speed passenger trains, each pair of originating/terminating EMU may be converted into 0.4 pair of originating/terminating ordinary speed passenger train; the EMU requiring immediate turning round or stopover may be respectively converted into 0.9 pair or 0.7 pair of originating/terminating EMU.

3 When there are 50 pairs or above originating/terminating ordinary speed passenger trains, the number of receiving-departures may be determined by analysis and calculation.

2 The number of receiving-departure tracks for high-speed railway station and intercity railway station shall be determined by analysis and calculation in accordance with train pair number, train operation plan, transportation organization mode, lead-in line number and the train headway of connecting line.

9.2 Layout Plan of Passenger Station

9.2.1 The layout plan of passenger station shall be determined in accordance with the nature and number of the lead-in lines, traffic capacity of the line, station operation workload, passenger train operation plan, station nature and operation requirement.

9.2.2 Through-type layout plan should be adopted for the passenger station, the mixed layout plan of through-type with partial dead-end type tracks may be adopted for the passenger station mainly for originating/terminating trains, and the dead-end layout plan may be adopted for the passenger station which is completely for originating/terminating trains and is located at the terminal end of the main line.

9.2.3 The layout plan of passenger station primarily with locomotive pulled passenger trains may adopt the layout of Figure 9.2.3-1, Figure 9.2.3-2 or Figure 9.2.3-3.

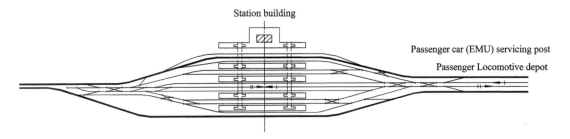

Figure 9.2.3-1 Layout Plan of Through-type Passenger Station with the Transfer Track of Depot Arranged between Two Main lines

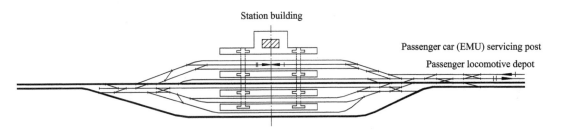

Figure 9.2.3-2 Layout Plan of Through-type Passenger Station with the Transfer Track of Depot Arranged at One Side of Main lines

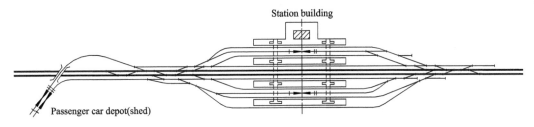

Figure 9.2.3-3 Layout Plan of Through-type Passenger Station with Main lines Arranged between Transfer Tracks of Depot

9.2.4 The layout plan of originating station for high-speed railway may be arranged as per Figure 9.2.4-1, Figure 9.2.4-2 or Figure 9.2.4-3.

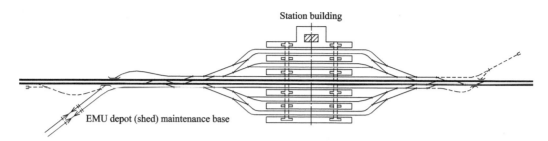

Figure 9.2.4-1　Layout Plan of Through-type Originating Station for High-speed Railway

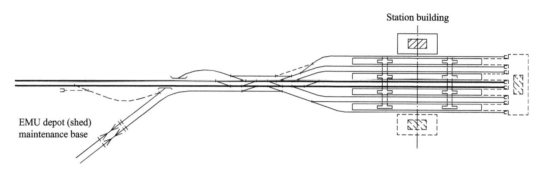

Figure 9.2.4-2　Layout Plan of Dead-end Type Originating Station for High-speed Railway

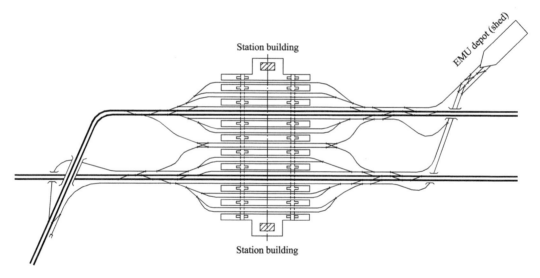

Figure 9.2.4-3　Layout Plan of Originating Station with the Intersection of Two High-speed Railways

9.2.5　The layout plan of originating station for intercity railway may be arranged as per Figure 9.2.5-1 or Figure 9.2.5-2.

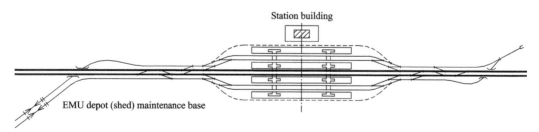

Figure 9.2.5-1　Layout Plan of Through-type Originating Station for Intercity Railway

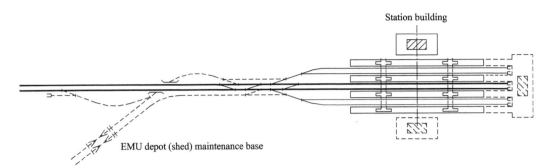

Figure 9.2.5-2 Layout Plan of Dead-end Type Originating Station for Intercity Railway

9.3 Passenger Facilities and Equipment

9.3.1 The layout of passenger station building shall comply with the following provisions:

1 The passenger station building shall be close to urban center district, the double-side station building may be adopted if located at urban center, and the dead-end type passenger station building may be located at the end of the platform.

2 Based on specific conditions, the passenger station building may be divided into parallel track-level type, above track-level type, under track-level type and multi-story grade separation type.

3 The passenger station building of super-large and large passenger station should be designed into an integrated transportation transfer terminal or urban complex together with urban rail transit station, long-distance bus station as well as bus stop, if conditions are available.

9.3.2 The layout of passenger platform shall comply with the following provisions:

1 The passenger platform should not be adjacent to the main line, especially, shall not be adjacent to the main line with freight trains passing through. If reconstruction is difficult, the lower platform adjacent to the main line may be remained.

2 The passenger platform of high-speed railway and intercity railway shall be arranged at the central position of the car yard, and the passenger platform of mixed traffic railway should be arranged at the central position of the car yard.

3 The platform should be arranged at the tangent section, and may be intruded into the curve section under difficult conditions.

4 When the platform is adjacent to the main line and the train passing speed is over 80 km/h, the platform safety marking shall be set 1.5 m away from the platform edge, and safety protection facilities shall be set 1.2 m away from the platform edge in case of need; when the train passing speed is lower than 80 km/h, the platform safety siding shall be set 1.0 m away from the platform edge.

5 When the platform is located at the side of the receiving-departure track, the platform safety marking shall be set 1.0 m away from the platform edge.

6 If platform screen door is set, the platform safety marking may not be set.

9.3.3 The design of passenger platform shall comply with the following provisions:

1 The passenger platform for high-speed railway shall be designed 450 m long. The length of the platform for intercity railway shall be calculated and decided according to train configuration length and parking deviation, it may be designed as 220 m long for 8 cars configuration. The platform for receiving and departing locomotive pulled passenger trains shall be designed as 550 m long, and it may be appropriately reduced based on the configuration of passenger trains under difficult conditions. The locomotive length and necessary entry/exit length shall be additionally considered in the design of the platform length for dead-end type passenger station under locomotive pulled condition.

2 The width of the passenger platform shall be determined according to the factors such as station nature, platform type, passenger flow density, luggage handling tools, safety evacuation distance, width of passageway entrance/exit on the platform, and the width of buildings(structures)on the platform. The width of the passenger platform for high-speed railway and mixed traffic railway may be selected from Table 9.3.3-1, and the width of the passenger platform for intercity railway may be selected from Table 9.3.3-2.

Table 9.3.3-1 Width of Passenger Platform for High-speed Railway and Mixed Traffic Railway(m)

Item	Super large station	Large station	Medium station	Small station
Distance from the protruding edge of station building or structure to the edge of main platform	20.0~25.0	15.0~20.0	8.0~15.0	8.0
Island intermediate platform	11.5~12.0	11.5~12.0	10.5~12.0	10.0~12.0
Side intermediate platform	8.5~9.0	8.5~9.0	7.5~9.0	7.0~9.0

Notes: 1 The width of main platform outside the range of station building and/or structure shall not be smaller than the width of side intermediate platform.
    2 The width of platform for mixed traffic railway should adopt the lower limit value.
    3 The width required for platform screen door is not considered in the values above.

Table 9.3.3-2 Width of Passenger Platform for Intercity Railway(m)

Item	Originating station	Intermediate station
Main platform or side intermediate platform(net width)	5.0	5.0
Island intermediate platform	11.5	8.5~10.5

Notes: 1 The distance from the protruding edge of station building and/or structure to the edge of main platform shall comply with the provisions of *Code for Design of Railway Passenger Station*(GB 50226).
    2 The width of the main platform or side intermediate platform measured at passenger passageway entrance/exit shall be widened based on the width of the passageway entrance/exit.
    3 For underground station or there are three or more entrance/exit on the platform, the width of island intermediate platform shall be calculated and decided based on the layout of the entrance and exit.
    4 The width of platform screen door and building(structure)is not considered in the values above.

3 The height of passenger platform: the passenger platform for high-speed railway and intercity railway shall be 1.25 m higher than rail surface; it should be 1.25 m higher than the rail surface at the side of adjacent receiving-departure track without stopover of out-of-gauge freight train for mixed traffic railway, and it shall be 0.3 m higher than the rail surface at the side of adjacent receiving-departure track with stopover of out-of-gauge freight train. The platform of 0.5 m higher than the rail surface in the existing station with only stopover of ordinary speed passenger trains may be remained.

4 When the platform is located at curve section, the minimum width of platform end should not be smaller than 5.0 m.

5 The transverse slope of the platform surface shall not be more than 1‰.

6 Both ends of high passenger platform shall be set with steps or ramp with protective fence, the fence door with width of not less than 1.0 m shall be set, and "NO ENTRY" signboard shall be erected. The location where fence is set shall comply with the provision of structural clearance.

9.3.4 The setting of platform screen door shall comply with the following provisions:

1 If the main line is adjacent to the platform, with passing or stopover of passenger trains, or the platform is used as waiting space, the platform shall be set with platform screen door.

2 The distance from the edge of platform screen door to the entrance/exit of passageway of passenger entering and exiting station or to the edge of building(structure) shall not be smaller than 2.5 m.

3 The distance between the platform screen door and the platform edge shall be comprehensively determined based on train operation mode, clearance requirement, train operation speed, the integrated control style of signal system, and the safety for passenger boarding & alighting.

9.3.5 The distance between the edge of platform entrance/exit and/or the building(structure) to the edge of passenger platform adjacent to the track shall not be less than 3.0 m, and it shall not be less than 2.5 m for medium and small station under difficult conditions. It shall not be less than 2.0 m for reconstruction of existing station at one side of the platform.

9.3.6 The setting for passageway of passenger entering and exiting station shall be determined according to traffic volume, design of passenger station building and the passenger entering and exiting station streamline, and it shall also comply with the following provisions:

1 Overpass or underpass passageway shall be adopted for passenger entering and exiting station, and the underpass shall be preferably selected. If the elevated overcrossing waiting room is adopted, the station entering overpass shall be combined with the elevated waiting room.

2 The passageway for passenger entering and exiting station for medium passenger station and above shall be set separately which shall be conveniently accessible for passengers, and the cross interference shall be reduced.

3 The number of the passageway for passenger entering and exiting station: one passageway shall be set for small passenger station, one entering passageway and one exiting passageway shall be set respectively for medium passenger station; the passageway for passenger exiting shall be overall considered for large and super-large passenger stations.

4 The minimum width of the passageway for passenger entering and exiting station shall comply with Table 9.3.6-1.

Table 9.3.6-1 Minimum Width of Passageway for Passenger Entering and Exiting Station(m)

Item	Passageway for passenger entering and exiting station		
	Super-large passenger station	Large passenger station (Originating station of intercity railway)	Medium and small passenger station (Intermediate station of intercity railway)
Minimum width	12	8~12	6~8

5 The net height of the underpass shall not be less than 2.5 m.

6 The passenger overpass and underpass should be set with two-way entrances and exits towards different platforms. The width of entrance and exit of passenger platform for high-speed railway and mixed traffic railway shall comply with Table 9.3.6-2, and the width of entrance and exit of the passageway for passenger entering and exiting station for intercity railway shall comply with Table 9.3.6-3; if escalators or elevators are set at the entrance and exit of passageway for passenger entering and exiting station, the width shall be widened according to the number and requirements of the lifting equipment.

Table 9.3.6-2 Width of Entrance and Exit of Passenger Platform for High-speed Railway and Mixed Traffic Railway(m)

Item	Super large and large station	Medium station	Small station
Main platform Island intermediate platform	5.0~5.5	4.0~5.0	3.5~4.0
Side intermediate platform	5.0	4.0	3.5~4.0

Note: The width of installation of one escalator is included in the above width of the passageway for passenger entering and exiting in super large and large station.

Table 9.3.6-3 Width of Entrance and Exit of Passageway for Passenger Entering and Exiting Station of Intercity Railway(m)

Originating station	Intermediate station
4.5~5.0	3.0~4.0

Note: The width of installation of one escalator is included in the above width of the passageway for passenger entering and exiting in the originating station.

7 Under the premise that the safety and the operation function being guaranteed, the existing passageway for passenger entering and exiting station may be expanded in the extension and upgrading of the existing passenger station.

9.3.7 The operation underpass towards the platform shall be set for the station, The arrangement of operation underpass shall comply with the following provisions:

1 Underpass number: at least one underpass shall be set for super large and large passenger station; one underpass may be set for the intermediate station with operation for originating and terminating passenger trains; and it may be combined with the passageway for passenger entering and exiting in small station where the separate operation underpass shall not be set.

2 The underpass width shall not be less than 5.2 m.

3 The net height of the underpass shall not be less than 3.0 m.

4 The underpass should be set with one-way entering and exiting passageway towards different platforms, which shall be set at the end part of the platform, and the width shall not be less than 4.5 m. If restricted by conditions and there is traffic indicator at the entering and exiting passageway, the width shall not be less than 3.5 m.

9.3.8 The passenger platform shall be covered with shelter, and the length of shelter shall be in line with the platform length. The shelter length of the small station shall be appropriately shortened according to passenger volume and local rainfall.

9.3.9 For the station with water supply and sewage discharging operation, the water outlet and sewage discharging facilities shall be set beside the corresponding receiving-departure track. The number of receiving-departure tracks equipped with water hydrant and sewage discharging facilities shall be determined according to related provisions.

9.4 Passenger Car Depot(Shed)

9.4.1 When passenger station is equipped with passenger car depot(shed), the mutual disposition shall meet the carrying capacity of the station, the cross interference of the throat shall be reduced, the running distance of locomotive, passenger train set and EMUs shall be reduced, and the arrangement shall be determined after overall comparison based on topographical, geological conditions, as well as urban planning.

1 Passenger car servicing depot and EMU depot(shed) shall be longitudinally arranged at the side of the main line outside of the throat area of passenger station with less receiving-departure trains.

2 For the passenger station with mainly locomotive pulled passenger trains, when all or most of the passenger trains shall have stopover operation, the passenger car servicing depot and locomotive equipment servicing facilities may be longitudinally arranged between the two main lines of the double-track railway.

3 If the pair number of originating and terminating passenger trains is small, and there is no potentially expansion and upgrading for the station in the long run, the passenger car servicing depot

may be transversely set with the passenger station.

9.4.2 The passenger car servicing depot and the passenger locomotive equipment servicing facilities of the passenger station should be combined so as to meet the operation requirement of leading locomotive for leaving and arriving.

9.4.3 For the passenger station with main operation for locomotive pulled passenger trains and with originating and terminating operation of EMUs, facilities shall be set for operation of EMU inspection, servicing and storage. The existing passenger car servicing facilities shall be fully utilized for EMU facilities, and EMU facilities shall be combined with the passenger car servicing depot. The operation of ordinary speed passenger trains shall be separated from that of the EMUs.

9.4.4 When the passenger station is longitudinally arranged with the passenger car servicing depot or EMU operation shed, the number of the connecting lines between the station and the depot(shed) shall be determined according to the number of servicing train sets entering and exiting the depot (operation point), times of locomotives entering and exiting depot, the layout plan of passenger car servicing depot or EMU depot(shed), shunting workload, as well as the distance between the station and the depot(shed). One connecting line shall be set between the station and the depot(shed), and two connecting liens may be set for insufficient capability. The connecting line between the station and the depot(shed) should be long enough for train set dispatching and car washing machine. If the car washing machine is installed between the passenger servicing yard and technical servicing yard of the passenger car servicing depot, the connecting line between the station and the depot may be appropriately shortened.

9.4.5 One shunting neck shall be set if the passenger station is transversely set with the passenger car servicing depot. If the passenger station is longitudinally set with the passenger car servicing depot, the connecting line between the station and the depot or the tracks of the passenger servicing yard may be used as the shunting neck; if there are many train sets entering and exiting the depot, and the connecting line between the station and the depot is of insufficient capability, 1~2 shunting necks shall be set. The effective length of the shunting neck shall not be less than the effective length of the receiving-departure track of passenger trains.

10 Railway Logistics Center

10.1 General Requirements

10.1.1 Railway logistics center shall take the national, local and railway development planning as the basis, coordinate with local economic development, industry distribution and freight source, and consider short-term investment benefits and long-term development demands as a whole, implementing one plan by stages.

10.1.2 Design of railway logistics center shall meet the requirement for containerized freight, mechanized loading and unloading, fast transport, automatic storage, integrated warehouse and distribution, integrated information, intelligent safety inspection & monitoring, convenient service and modern management.

10.1.3 Railway logistics center shall include the operating car yard, logistics functional zone and other logistics auxiliary facilities, the operating car yard and the logistics functional zone should be concentrated.

10.1.4 Railway logistics center may be divided into comprehensive railway logistics center and special-purpose logistics center according to the freight category and nature. Comprehensive railway logistics center may be divided into level I, level II and level III centers according to throughout, service function, construction scale and the role in the network. Those comprehensive railway logistics center may also be divided into freight service type, production service type, trade service type, port service type and comprehensive service type railway logistics center according to service type.

10.1.5 Comprehensive railway logistics center may be composed of trade and exhibition zone and functional zones such as container, heavy and bulky goods, packaged goods, commodity cars, stacked freight, storage and delivery, hazardous goods, refrigeration, circulation and processing.

10.1.6 Special railway logistics center shall have large loading/unloading point for bulk cargo and special cargo, and it may be divided into railway logistics centers for heavy and bulky cargo, stacked cargo, dangerous cargo, commodity car and self-running dumper according to cargo category.

10.1.7 The special railway logistics center for danger cargo should be separately established to transport, store, handle and distribute dangerous cargo. The functional zone for dangerous cargo shall be separately divided in the comprehensive railway logistics center. The location and equipment of the special railway logistics center for danger cargo and the separately divided functional zone for dangerous cargo within the comprehensive railway logistics center shall comply with current national norms on fire prevention, explosion protection, anti-poison, health as well as environmental protection.

10.1.8 In case of requirements on military logistics, railway logistics center may be equipped with military transport and military supply facilities and shall comply with standards for railway military facilities.

10.2 General Layout

10.2.1 Function of the railway logistics center shall use differential design and modular design according to cargo category and customer type, the functional modules shall be coordinated for easy logistics operation.

10.2.2 The streamline design of the railway logistics center shall comply with the following provisions:

1 The total distance for transport and delivery should be possibly short.

2 The streamline of the transport equipment should be short and straight.

3 The streamline of railway vehicle shall be easy for train direct arriving and direct departure, and shunting.

4 The streamline of road vehicle should be designed in unidirectional circular type, the gate area and densely operated work area shall be set with buffer zone.

5 The streamline of personnel operation shall be smooth and safe, the passageway for personnel only shall be set among the crowded buildings.

10.2.3 The layout of the operation car yard and the loading/unloading yard of railway logistics center shall be determined after comparison in economy and technology based on workload, cargo category, operation nature and local conditions. The layout shall be possibly compact, the land shall be fully used and conditions for long-term development shall be reserved. Layout design may comply with the following provisions:

1 The operation car yard and the loading/unloading yard may use transverse or longitudinal type.

2 The layout plan of the loading/unloading yard may use through type, dead-end type or mixed type; the layout of the track may use parallel type, partial parallel type or radiant type.

3 The layout plan of container loading/unloading yard should use transverse through type, longitudinal through type, transverse mixed type, transverse dead-end type or other layout plans; the loading/unloading yard for packaged cargo, heavy and bulky cargo and bulk cargo should use transverse-type layout plan; the loading/unloading yard of bulk cargo quick delivery station should use the transverse through type layout plan of receiving train, three platforms clamped with two tracks; the loading/unloading yard for commodity car and dangerous cargo should use dead-end type layout plan.

10.2.4 The layout of different functional zones of railway logistics center shall comply with the following provisions:

1 The functional zone for heavy and bulky cargo should be close to the container functional zone.

2 The commodity car should be close to the container functional zone, the road and gate accessing the functional zone of the commodity car should be separately established.

3 The functional zone of railway bulk cargo with flowing dust pollution should be separately set, and shall be placed outside the railway logistics center and at the downwind area of prevailing wind direction if it is required to combine with other functional zones, and should be far away from the functional zones of packaged cargo, refrigeration and commodity car.

4 The functional zone of dangerous cargo shall be far away from other functional zones, production, office and living facilities, and shall be at the downwind area of prevailing wind direction.

5 The functional zone of storage and distribution and the production functional zone where railway handling is not required should be far away from railing handling line, and shall be arranged close to the gate area, and the separate door shall be set if necessary.

6 The business area, trading zone, and exhibition area should be arranged on the main passageway next to the railway logistics center, and shall be separated with the handling operation area and storage operation area. The comprehensive servicing building and the public parking area shall be arranged close to the gate area, the parking area for passenger trains and freight trains should be established separately.

7 The customs supervision operation area should be concentrated established and shall be in closed management.

10.3 Operation Car Yard

10.3.1 The receiving-departure yard and the shunting yard of the railway logistics center should use transverse-type layout plan, the throat area shall be set with the parallel route for train arriving and departure as well as for train shunting.

10.3.2 The number of receiving-departure track shall be calculated and decided according to traffic volume, train nature and technical operation process. When the loading/unloading track is with receiving-departure conditions or the train technical operation is not required for receiving-departure track, the number of receiving-departure track may be reduced. The effective length of the receiving-departure track shall be the same as that of the receiving-departure track for the main line railway station; the effective length of the receiving-departure track for the district transfer train may be determined by adding 30 m additional braking distance to the length of the district transfer train.

10.3.3 The number of the shunting track shall be determined according to the number of loading/unloading points, number of operation cars, shunting operation ways as well as the transport characterization of special freight. The effective length of the shunting track should be the same as that of the receiving-departure track, and it may be shortened under difficult topographical condition or for large rebuilding and expansion project workload.

10.3.4 The stabling siding shall be established according to the pair number of originating and destination trains, freight operation volume, complexity of car taking-out and placing-in operation, distance of the operation car yard and logistics functional zones, and the work capacity of equipment at logistics functional zones. The stabling siding may be set in the operation car yard or the logistics functional zone.

10.3.5 The operation car yard shall be set with shunting neck whose number shall be determined according to the plane layout and the scale of the operation car yard, and the effective length shall not be less than that of the loading/unloading track. If the siding connected to the operation car yard is of sufficient capacity and the shunting requirement can be met, the siding may be used as the shunting neck.

10.3.6 If the operation car yard is busy and quite far from locomotive servicing or car maintenance point, the locomotive servicing or station repair facilities may be established separately.

10.4 Freight Traffic Facilities and Equipment

10.4.1 Railway logistics center shall be equipped with facilities for consignment, weighing, ticket printing, fare collection, information and delivery, and shall also be equipped with facilities for freight platform, warehouse and goods allocation, stacking yard, container handling place, shelter, water drainage, fire fighting, lighting, road and enclosing wall, freight security detection and protection, centralized video surveillance and IT application system as required. The place for freight car washing and pollution cleaning shall be equipped with pollution disposal and sewage facilities. Freight handling operation shall use mechanical equipment.

10.4.2 Railway logistics center mainly for containers shall be equipped with facilities as per the requirement of whole train.

10.4.3 The effective handling length of goods loading/unloading track as well as the length of freight storage warehouse shall be determined according to freight volume, average net load of various freight vehicle, freight stacking volume of unit area, time of freight occupying cargo space, times of daily car taking-out and placing-in, number of goods rows and the width of storage space per row, etc.

10.4.4 The distance between goods loading/unloading track and the adjacent track shall be calculated and decided according to type of handling machinery, arrangement of storage space, road as well as the operation nature of the adjacent lines.

10.4.5 The layout of platform and the loading/unloading track may use the layout plan of one platform with one track or one track between two platforms according to the receiving-departure volume, transportation organization and requirement of handling operation, or may use other layout plans like two platform clamping two tracks, two platforms clamping three tracks and three platforms clamping two tracks.

10.4.6 The freight platform may be divided into general freight platform, end platform, high platform and low platform according to its purpose and the distance to rail surface, the design of the platform shall comply with the following provisions:

    1 The length and width of freight platform shall be determined according to freight category, work volume, number of car taking-out and placing-in, process of handling operation, type of handling machinery and warehouse layout.

    2 The top surface of the edge for general freight platform should be 0.95~1.1 m higher than the rail surface at the side of the track, and should be 1.1~1.3 m higher than the station yard at the side of the site, and the height adjustable lift platform shall be equipped according to actual need.

    3 Water shall be drainage outward for freight platform, and the gradient should be 1%.

    4 The general freight platform next to the loading/unloading track should use rectangular layout, or L type, convex type or concave type platform which is connected with the common platform or end platform according to operation requirement.

    5 The end part of general freight platform shall be set with slope, convenient for forklift operation, the slope width shall not be less than 3.5 m, the slope gradient shall not be more than 1 : 12, and the slope gradient of the platform for commodity car handling shall not be more than 1 : 7.

    6 For commodity car handling, the adjustable double-layer loading/unloading platform or dead-end type freight platform may be used together with mobile commodity car handling ladder. The projected length of the upper layer for the adjustable double-layer handling platform should be 49 m, the width of single passage shall not be less than 5 m; the lower layer shall be designed as per dead-end type freight platform, the platform length shall not be less than 6 m, and the width of single passage shall not be less than 5 m. The length of the dead-end type freight platform for auxiliary mobile commodity car handling ladder shall not be less than 10 m, and the width of single passage shall not be less than 5 m.

    7 The warehouse platform shall use monolithic casting, with platform edge inlaid with angle steel.

10.4.7 The freight warehouse or cargo shed shall be designed according to freight type and storage space, and shall comply with the following provisions:

    1 The freight warehouse or cargo shed should use rectangular layout, and the warehouse or cargo shed with huge work volume or in rainy or snowy areas should use cross-track layout.

    2 The shelter shall be set for warehouse or cargo shed at site side or railway side, and the shelter shall meet the requirement for operation in snowy or rainy weather.

    3 The width and length of the warehouse or cargo shed may be determined according to annual freight volume, location, size of storage space, layout of storage space, freight stacking height, type of handling machinery, operation style and operation category. The width and length of storage and distribution type or storage type warehouse or cargo shed shall be determined according to logistics service requirement, and should not be less than 30 m; the transshipment warehouse or cargo shed should be 18 m or above. The width and length of warehouse or cargo shed for fast freight sorting shall be determined according to the sorting equipment and freight sorting volume. The length of the warehouse or cargo shed beside the loading/unloading track shall be in accordance with the effective length of the loading/unloading track.

  4 The height of warehouse or cargo shed shall be determined according to freight stacking height, requirement of handling machinery passage and requirement of staking operation. The net height of single-layer warehouse or cargo shed with storage rack should not be less than 8.5 m, and the net height of warehouse or cargo shed with freight storage as well as the warehouse or cargo shed with less occupied freight racks should not be higher than 7 m; the bottom net height of multiple layer warehouse should not be less than 6 m, and the clear height of storey should not be less than 5 m; the height of warehouse or cargo shed equipped with bridge crane shall be comprehensively decided according to the minimum lifting height of freight and the device height.

  5 The spacing of the axis for warehouse outer wall with the platform edge shall be calculated and decided according to type of operation machinery, turning radius and the freight category in the warehouse, it should be 4 m at the side of the railway, and it should not be less than 4 m at the side of the site.

  6 The width of handling operation site at the side of platform warehouse site shall meet the requirement for highway freight vehicle, and generally should not be less than 30 m.

10.4.8 Design of road and operation site for railway logistics center shall comply with the following provisions:

  1 The road width and turning radius shall meet the requirement of handling machinery and highway freight vehicle, and the loop should be formed.

  2 The work passageway and site shall be hardened cement, and corresponding road standard shall be used.

  3 Parking place may be established inside the site or outside the site according to actual need, the buffer zone shall be established for highway vehicle outside the entrance of the operation zone. The parking place inside the site shall meet the parking requirement of auxiliary work vehicle and mobile machinery, and the parking place outside the site shall meet the temporary parking requirement of commercial vehicle and the vehicle to be handled.

11 Hump

11.1 General Requirements

11.1.1 The hump shall be divided into large-capacity one, medium-capacity one and small-capacity one according to daily breaking up capacity, and their equipment shall be equipped in compliance with the following provisions:

1 For large-capacity hump with daily breaking up capacity over 4 000 cars, 30 or more shunting tracks and two rolling tracks shall be set, and the hump route control system, speed control system for rolling train set, speed control system for humping part, as well as the halting and preventing car rolling control system at the end of the shunting yard shall be equipped.

2 For medium capacity hump with daily breaking up capacity of 2 000~4 000 cars, 17~29 shunting tracks and two rolling tracks shall be set, and the hump route control system, speed control system for rolling train set, speed control system for humping part, as well as the halting and preventing car rolling control system at the end of the shunting yard shall be equipped.

3 For small capacity hump with daily breaking up capacity of lower than 2 000 cars, 16 or less shunting tracks and one rolling track shall be set, and the hump route control system, and the halting and preventing car rolling control system at the end of the shunting yard shall be equipped, and the hump locomotive signal should be also equipped.

11.1.2 Hump type and technical equipment shall be determined according to short-term breaking up volume in hump design, and conditions for long-term development shall be reserved based on traffic volume increasing and technical equipment condition. If phased transition project is too complicated, the design plan for phased transition shall be worked out. For the technical upgrading of the exiting hump, the line plan and profile shall be upgraded in combination with the used speed regulation system.

11.2 Plan of Hump Track

11.2.1 The track plan of hump rolling part shall comply with the following provisions:

1 The track group layout shall be used, the shunting track number for every track group should be 6~8; No. 6 symmetrical turnout shall be used, or other symmetrical turnouts shall be used under difficult conditions; when it is especially difficult for line connection outside of shunting yard, No. 9 single turnout may be used. For small capacity hump with less number of shunting track, if it is difficult to use No. 6 symmetrical turnout, No. 9 lateral turnout or multiple ladder track plan may be used. If it is very difficult for reconstruction, the original ladder track plan may be retained.

2 The curve radius should not be less than 200 m, and 180 m may be used under difficult conditions.

3 The curve may be directly connected with the stock rail or frog heel of the turnout (except the front part of No. 1 shunt turnout), the track gauge widening and outer rail super elevation may be treated within the curve range.

11.2.2 The minimum spacing between the hump crest and the stock rail of No. 1 shunt turnout shall be 30~40 m. If there are turnouts between the hump crest and the No. 1 shunt turnout, the spacing may be determined according to specific circumstances.

11.2.3 When the receiving yard is set in front of the hump, 2 pushing tracks shall be set, if double humping operation style is used, 3~4 pushing tracks may be set; when the receiving yard is not set in front of the hump, 1~2 pushing tracks may be set according to the breaking-up volume. The spacing of pushing tracks shall not be less than 6.5 m. The pushing track where release-coupler is operated frequently shall be designed as straight line, and symmetrical turnouts should not be used close to hump crest. The line of sight where release-coupler is operated frequently shall meet the requirement of observation for shunting personnel, and the work passage shall meet the requirement of security for shunting personnel.

11.2.4 The hump with 2 pushing tracks and the track group is 4 or above shall be set with 2 rolling tracks.

11.2.5 Large capacity and medium capacity hump should be set with 2 prohibitive humping track; the effective length may be 150 m. It may be combined with detour track into one no-humping car storage if there are less prohibitive humping cars. The prohibitive humping car storage may be set according to actual need for small capacity hump. If the prohibitive humping track separates from the pushing track, No. 9 single turnout shall be used, the frog shall be set on the platform of the hump crest. The prohibitive humping track shall be set away from the buildings like signal cabin, and the parking car on the prohibitive humping track shall not hinder the observation of shunting personnel.

11.2.6 If the receiving yard is set in front of the hump, it shall bypass the hump crest and the detour track of car retarder; if the receiving yard is not set in front of the hump, the detour track may be set according to actual need.

11.2.7 For the design of hump track plan, the protection section required by car retarder and centralized control turnout shall be reasonably arranged, and the location of buildings shall be considered according to operation requirement, like hump signal building, hump crest connector room, shunting operator room and the power house of car retarder.

11.3 Profile of Hump Track

11.3.1 For large capacity hump, medium capacity hump and the small capacity hump with speed regulation device at rolling section, the crest height shall ensure the hard rolling car to roll to the calculating point of the hard rolling track under adverse condition when the train set is broken up at a pushing speed of 5 km/h. The location of the calculating point shall be determined according to the hump speed regulating system. The crest height of the small capacity hump which is not installed with speed regulation device at rolling section shall ensure that the easy rolling car to roll to the fouling post of the easy rolling track in the shunting yard is not less than 18 km/h under favorable car rolling condition when the train set is broken up at a pushing speed of 5 km/h; when the shunting track is equipped with car retarder, the speed at which the easy rolling cars roll to where car retarder is located shall not exceed the speed allowed by their braking power.

11.3.2 The profile of the track for hump rolling section shall be designed as downhill towards the shunting yard, and its gradient composition shall comply with the following requirements:

1 Accelerating grade shall not exceed 55‰, and shall not be less than less than 35‰ under difficult conditions.

2 Intermediate grade may be designed as multiple grades or one-section grade. The gradient of the section with car retarder should not be less than 8‰. The point of gradient change for intermediate grade and accelerating grade should be set in front of the stock rail of the first branching turnout.

3 Gradient within the switching area: average gradient should not be more than 2.5‰, and the edge track group shall not be more than 3.5‰.

4 The profile of the track for hump rolling section shall be calculated according to the used speed regulation system and shall comply with the following requirements:

    1) Sufficient spacing shall be ensured under adverse car rolling condition and when the single car is rolled continuously difficult-medium-difficult through car retarder, different branching turnout and fouling post at 5 km/h pushing speed. When the hump retarders speed control system and station track retarders speed control system is used, the hard rolling train group-single easy rolling car shall be calculated.

    2) The speed shall not exceed the rated value for car to enter the speed control device.

    3) The speed for car passing through different branching turnout shall not exceed the speed used for protecting the section length.

11.3.3 The track profile of hump pushing section shall ensure that under difficult conditions, one shunting locomotive may start the train set. The coupler compressing grade shall be set in front of the hump crest, the gradient shall not be less than 10‰ and the length shall not be less than 50 m.

11.3.4 For the vertical curve radius of different grade section connecting hump tracks, the coupler compressing grade at the hump crest shall not be less than 350 m; the adjacent accelerating grade shall be 350 m; other rolling section and detour track shall not be less than 250 m and 1 500 m respectively.

11.3.5 The net platform length at crest should be 7.5~10 m.

11.3.6 The profile of prohibitive humping track shall be concave shape, from the starting turnout to the fouling post shall be designed as downhill slope, the middle parking section should be designed as flat slope, within the range 10 m to bumper post shall be designed as 10‰ uphill slope.

11.4 Other Requirements

11.4.1 The braking capacity of hump velocity modulation equipment is decided by computers, and the safety level will be added according to equipment requirement. The velocity modulation equipment of the rolling part for large capacity and medium capacity humps shall use car retarder. The hump which is connected with 16 or above shunting tracks should be set with two-level or one-level spacing braking position. When two-level braking position is equipped, the total braking capacity shall ensure the easy rolling cars to roll to the fouling post of the easy rolling track at the speed lower than 5 km/h under favorable conditions, and the car is broken up at a pushing speed of 7 km/h, as well as the cars are all braked after the first and the second spacing braking positions.

11.4.2 The car retarder shall be set on the straight line, there shall be at least 14 m straight line section in front and the car retarder within the shunting track, and the effective velocity hump crest of retarder should not be less than 1.3 m.

11.4.3 The starting/ending point of the vertical curve for the grade change point of crest peak and rolling part, as well as the main grade change points within the shunting track shall be set with line horizontal stakes.

11.4.4 The setting of related equipment for hump and production buildings shall comply with the following requirements:

    1 The operation safety and observation of hump shunting personnel shall not be hindered.

    2 The hump signal building shall be set in a centralized way, its location and the connector room at crest peak shall be at the same side with the operation point of the main lifter, and the position shall ensure the operators to clearly observe the car running condition at the hump crest peak, rolling part and the car retarder at the spacing braking position.

12 Industrial Station and Harbour Station

12.1 General Requirements

12.1.1 The industrial and mining enterprises, the industrial district or the port with a large number of loading/unloading operation may be set with railway industrial station or harbour station according to actual need.

12.1.2 The number of industrial stations serving the same enterprise or industrial district shall be determined according to enterprise property, scale of production, production process, layout of enterprise or industrial district, source of raw materials, product flow direction, location of enterprise or industrial district and their mutual relation with railways.

12.1.3 The position of the industrial station and harbour stations shall be selected according to the following requirements:

   1 According to the location and general layout of enterprise and port, the traffic volume of railway and the receiving and delivery system, the station shall be constructed close to the railway or close to the enterprise or the port, and the connection with the railway shall ensure the main traffic flow to be straight and smooth.

   2 The location of industrial stations and harbour stations shall be convenient for different operation station, partition car yard and taking out and placing in of cars in loading/unloading points.

   3 To coordinate with urban planning, to give consideration to local passenger and freight transportation, to meet the requirement of environmental protection, fire fighting and sanitation, and to be easy to connect, coordinate with and combined transport with other transportation modes.

12.1.4 The scale of industrial station and harbour station shall be determined according to the traffic volume of the enterprise or the port through the railway, transportation property, operation volume, management and delivery-receiving mode as well as the role of the station in the network. The general layout shall be worked out based on the planning of the enterprise or the port in station design period and the short term engineering shall be determined according to the principle of construction by stages.

12.1.5 The management, delivery-receiving mode and delivery-receiving position of the industrial siding shall be determined after technical and economic comparison and negotiation with the enterprise or the port based on specific condition.

12.2 Layout Plan of Industrial Station and Harbour Station

12.2.1 When the wagon delivery-receiving mode is used, the industrial station and harbour station undertake little transit shipment operation volume, close to enterprise station or harbour station and the topographical condition is suitable, the industrial station should be combined with the enterprise station or the harbour station should be combined with the port station; otherwise, they should be built separately.

12.2.2 The layout plan of the industrial station or the harbour station shall be determined according to hand-over style, operation volume, operation nature, the division of work for the station in the network, and the goods loading/unloading position. The design of the layout plan shall be selected based on the following provisions:

   1 When the goods delivery-receiving mode is used, the transverse-type layout plan (Figure

12.2.2-1) should be applied.

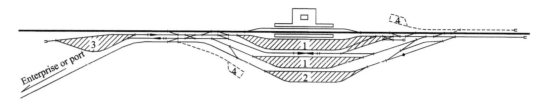

Figure 12.2.2-1 Transverse-type Layout Plan When Using Goods Delivery-Receiving Mode
1—Railway receiving-departure yard; 2—Railway shunting yard; 3—Railway locomotive depot; 4—Logistics center (proposed)

2 When the wagons delivery-receiving mode is used and the stations for both parties are built separately, the transverse-type layout plan (Figure 12.2.2-2) should be applied. If operation volume is huge, other suitable layout plans may be applied.

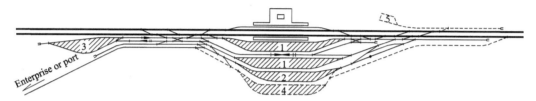

Figure 12.2.2-2 Transverse-type Layout Plan When Stations Being Built Separately for Both Parties
1—Railway receiving-departure yard; 2—Railway shunting yard; 3—Railway locomotive depot;
4—Delivery-receiving yard; 5—Logistics center

3 When the wagon delivery-receiving mode is used and the station is built jointly by both parties, the same type of layout plan of car yards should be used for both parties, either the transverse-type layout plan (Figure 12.2.2-3) or the longitudinal-type layout plan (Figure 12.2.2-4). If operation volume is huge, the bidirectional hybrid-type layout plan (Figure 12.2.2-5) or other suitable layout plans may be used under the condition that the station is built jointly by both parties.

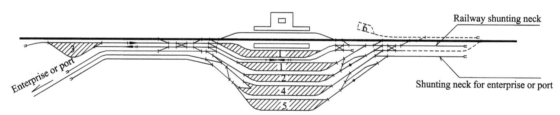

Figure 12.2.2-3 Transverse-type Layout Plan When Station Being Built Jointly by Both Parties
1—Railway receiving-departure yard; 2—Railway shunting yard; 3—Railway locomotive depot; 4—Arriving-depature yard and delivery-receiving yard for enterprise or port; 5—Shunting yard for enterprise or port; 6—Logistics center

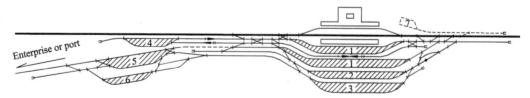

Figure 12.2.2-4 Longitudinal-type Layout Plan When Station Being Built Jointly by Both Parties
1—Railway receiving-departure yard; 2—Railway shunting yard; 3—Delivery-receiving yard; 4—Railway locomotive depot;
5—Arriving-depature yard for enterprise or port; 6—Shunting yard for enterprise or port; 7—Logistics center

Note: When yard 5 is also used as delivery-receiving yard, the connecting line represented by dash-line in the figure will be used, and yard 3 and its connecting line will be cancelled.

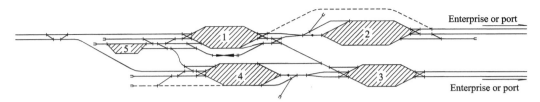

Figure 12.2.2-5 Bidirectional Hybrid-type Layout Plan When Station Being Built Jointly by Both Parties
1—Railway receiving-departure yard; 2—Marshalling-departure yard for enterprise or port; 3—Receiving yard for enterprise or port;
4—Railway marshalling-departure yard; 5—Railway locomotive depot

4 When the loading/unloading facilities are set in the industrial station or harbour station for enterprise or port to load and unload bulk cargo, the station layout plan shall be determined based on operation style and topographical condition.

12.3 Main Facilities and Equipment

12.3.1 If the wagon delivery-receiving mode is used and the industrial station or the harbour station containing delivery-receiving yard is separated from the corresponding station of the opposite party, the delivery-receiving yard of transverse-type layout should be built at one end or outside of the shunting yard; if other type of layout is applied, the delivery-receiving yard should be built at the side of the shunting yard. If the industrial station or the harbour station is jointly built by both parties and the transverse-type layout is used, the delivery-receiving yard shall be built between the car yards of both parties.

12.3.2 The delivery-receiving operation position shall be determined respectively according to the delivery-receiving mode used, the management style of industrial siding and the station layout.

1 If the goods delivery-receiving mode is used, the goods delivery-receiving operation passing in and out of the enterprise or the port may be handled on the loading/unloading track in the enterprise or the port. When mechanical loading/unloading facilities are equipped at the industrial station or harbour station by the enterprise or the port, the goods for loading should be delivered and handled at the loading track; the goods for unloading should be delivered and handled at the car yard or the unloading track in front of the unloading equipment.

2 If the wagon delivery-receiving mode is used and the stations of both parties are built separately, the delivery-receiving operation should be handled at the delivery-receiving yard which is set in the industrial station or the harbour station; when the industrial siding between stations of both parties is managed by the railway corporations, the delivery-receiving yard shall not be set at the industrial station or harbour station, and the delivery-receiving operation should be handled at the receiving-departure yard of the enterprise station or port station.

3 If the wagon delivery-receiving mode is used and the stations of both parties are jointly set, the delivery-receiving position should be determined according to the following condition:

1) If transverse-type or longitudinal-type layout is used, the delivery-receiving operation should be handled at the delivery-receiving yard; if delivery-receiving yard is not set, the delivery-receiving operation should be handled at the receiving-departure yard of the enterprise of the port.

2) If bidirectional hybrid-type layout is used, the delivery-receiving yard may not be set, the delivery-receiving operation may be handled at the receiving yards of both parties; if conditions are available, the delivery-receiving operation shall be handled at the receiving yard of the opposite party.

12.3.3 If industrial siding is connected at the industrial station or harbour station, it shall not interfere with the railway operation and station operation, it shall facilitate car taking-out and placing-in, as well as through transportation for enterprise and port, the track connecting point shall be set at the opposite side of the industrial station and harbour station where there is large traffic flow. If there are many industrial sidings that need to be connected, the overall planning shall be made and they shall be intensively combined and led to the same side of the car yard of the industrial station or harbour station. The specific rail connecting position shall be determined according to the following provisions:

1 If the goods delivery-receiving mode is used and there are whole train receiving-departure operations, it shall be connected with the receiving-departure yard, it shall be connected with the marshalling-departure yard if there is large amount of breaking-up and marshalling operation, and it may be connected with the shunting track, secondary shunting neck or other station track if traffic volume is small.

2 If the wagon delivery-receiving mode is used, the connecting position shall comply with the following provisions:

1) For the transverse-type industrial station or harbour station which is separated from the corresponding station of the opposite party, if the delivery-receiving yard is set, it shall be connected at the delivery-receiving yard, and the conditions shall be available for connection with other car yards; when the industrial siding between the stations of both parties is managed by railway corporation, it should be connected at the shunting yard; and it should be connected at the receiving-departure yard if there is receiving-departure operation for the whole train.

2) For the transverse-type industrial station or harbour station which is built jointly by both parties, it shall be connected at receiving-departure track and the delivery-receiving yard of the enterprise station or the port station, and the conditions shall be available for connection with other car yards. When the stations of both parties are combined into bidirectional hybrid-type layout, if the industrial siding passes in the enterprise or the port, it shall be connected at the marshalling-departure yard of the enterprise station or the port station; if the industrial siding starts from the enterprise or port, when the delivery-receiving position is located at the receiving yard of their own, it shall be connected with the receiving yard of the enterprise station or the port station; when the delivery-receiving position is located at the receiving yard of the opposite party, it shall be connected with the receiving yard of the railway.

12.4 Number and Effective Length of Station Tracks

12.4.1 The number of receiving-departure tracks for the industrial station or harbour station shall be determined according to the pair number of railway trains, times of arriving and departure as well as times of car taking-out and placing-in for enterprise station or port station trains, and the unified technical operation process of railway and factory (mine and port). The effective length of the receiving-departure track shall be in line with the effective length of the receiving-departure track of the connected railway. The effective length of the receiving-departure track only for arriving and departure of district transfer trains may be determined according to actual need.

12.4.2 The number and effective length of the shunting tracks for the industrial station and harbour station used for concentration and sending to network traffic flow shall be determined according to the group number stipulated by train marshalling plan, and the traffic flow and traffic flow nature of every day and night for every group number.

12.4.3 The number and effective length of the shunting track for the industrial station and harbour station used for concentration and sending to enterprise or port traffic flow shall comply with the following requirements:

1 If the goods delivery-receiving mode is used, it should be determined according to the number of operation stations, partition car yards and loading/unloading points of enterprise station or port station, number of cars sending to every operation station, partition car yard and loading/unloading points of every day and night, and the unified technical operation process of railway and factory (mine and port).

2 If the wagon delivery-receiving mode is used, when the industrial station is separately set with the enterprise station or the harbour station is separately set with the port station and the delivery-receiving yard is not set at the industrial station or harbour station, the number of shunting tracks should be determined according to the traffic flow which is concentrated at the shunting track and sent to the enterprise or the port, and the unified technical operation process of railway and factory (mine and port). When the delivery-receiving yard is set at the industrial station or the harbour station and set at the side of the railway shunting yard, the car which is sent to the enterprise or the port after breaking up should directly roll to the delivery-receiving yard, the shunting track used for concentrating and sending cars to the traffic flow of the enterprise or the port may not be set.

3 For the industrial station and harbour station with reserved cars, the storing track for spare cars shall be set appropriately according to the number of spare cars.

4 The effective length of the shunting track shall be determined according to the length of the train sending to the enterprise or the port.

12.4.4 The number and effective length of the delivery-receiving track shall comply with the following requirements:

1 When the industrial station is separately set with the enterprise station or the harbour station is separately set with the port station and the delivery-receiving yard is set at the industrial station or harbour station, the number of delivery-receiving tracks shall be determined according to the number of car handled every day and night, the times of car taking-out and placing-in of the delivery-receiving yard and the time of car delivery-receiving operation.

2 The effective length of the delivery-receiving track shall be in line with the effective length of the receiving-departure track of the industrial station or the harbour station. If there is big length difference of trains sending to enterprise or port with the trains sending to the railway, the effective length of some junction lines may be appropriately reduced but shall not be shorter than the effective length of the receiving-departure track of the enterprise station or the port station.

13 Border Station

13.1 General Requirements

13.1.1 The border station may be divided into the border station with the same track gauge and the border station with different track gauge based on the railway track gauge to be connected. The railway border station shall be set near the national border.

13.1.2 The general layout of the border station shall be designed according to passenger and freight traffic volume and based on topographical condition, geological condition and planning. The layout of the border station shall:

1 meet the capacity required and to meet the requirement of quick customs clearance.

2 follow the principle of planning once and construction in stages, the further development condition shall be reserved.

3 have the streamline of the station to meet the requirement of transshipment for the whole train set.

4 have a clearly defined transshipment zone.

5 reduce crossing of routes and operation interference, and to avoid the crossing of operation for different track gauges.

6 reduce the running distance of trains and locomotives.

7 use modern technical equipment.

8 possess the conditions to become the(international)logistics center.

13.2 Layout Plan of Border Station

13.2.1 According to the relative location relationship of the receiving-departure yard of different track gauge, the shunting yard and the transshipment yard, the border station shall use three layout plans, i. e, transverse type, longitudinal type and hybrid type.

1 If the traffic volume is small and it is restricted by the topographical condition, the transverse-type layout plan with the receiving-departure yard, the shunting yard and the transshipment yard of different track gauges may be used(Figure 13.2.1-1).

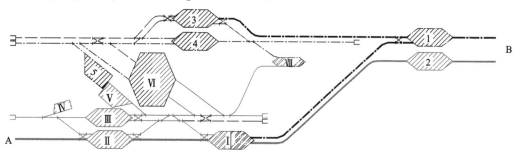

Legend: ——— Standard gauge track   —·—·— Non-standard gauge track

Figure 13.2.1-1 Transverse-type Layout Plan of Border Station

Ⅰ—Passenger station(and wheelset replacement workshop); Ⅱ—Receiving-departure yard of standard gauge; Ⅲ—Shunting yard of standard gauge; Ⅳ—Freight car repair point at station; Ⅴ—Logistics center of standard gauge; Ⅵ—Transshipment yard; Ⅶ—Locomotive depot; 1,2—Border check yard; 3—Receiving-departure yard of non-standard gauge; 4—Shunting yard of non-standard gauge; 5—Logistics center of non-standard gauge

2 If the traffic volume is big or it is not restricted by the topographical condition, the longitudinal-type layout plan with the receiving-departure yard, the shunting yard and the transshipment yard of two types of track gauges may be used(Figure 13.2.1-2).

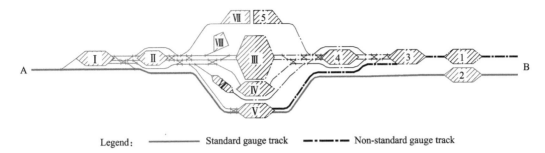

Legend: ——— Standard gauge track   —·—·— Non-standard gauge track

Figure 13.2.1-2   Longitudinal-type Layout Plan of Border Station

Ⅰ—Receiving-departure yard of standard gauge; Ⅱ—Shunting yard of standard gauge; Ⅲ—Transshipment yard; Ⅳ—Wheelset replacement yard; Ⅴ—Passenger station (and wheelset replacement workshop); Ⅵ—Locomotive depot; Ⅶ—Logistics center of standard gauge; Ⅷ—Freight car repair point at station;1,2—Border check yard;3—Receiving-departure yard of non-standard gauge; 4—Shunting yard of non-standard gauge;5—Logistics center of non-standard gauge

3 If the traffic volume is big or it is restricted by the topographical condition to some extent, the hybrid type layout plan shall be used with the receiving-departure yard and the shunting yard transversely arranged and then longitudinally arranged with the transshipment yard(Figure 13.2.1-3).

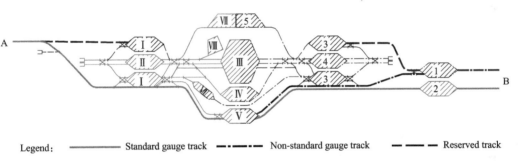

Legend: ——— Standard gauge track   —·—·— Non-standard gauge track   — — — Reserved track

Figure 13.2.1-3   Hybrid-type Layout Plan of Border Station

Ⅰ—Receiving-departure yard of standard gauge; Ⅱ—Shunting yard of standard gauge; Ⅲ—Transshipment yard; Ⅳ—Wheelset replacement yard; Ⅴ—Passenger station (and wheelset replacement workshop); Ⅵ—Locomotive depot; Ⅶ—Logistics center of standard gauge; Ⅷ—Freight car repair point at station;1,2—Border check yard;3—Receiving-departure yard of non-standard gauge; 4—Shunting yard of non-standard gauge;5—Logistics center of non-standard gauge

4 For the border station that is reconstructed or restricted by topographical conditions, other suitable layout plans may also be used.

13.3 Main Facilities and Equipment

13.3.1   The border station generally shall be set with the receiving-departure yard for passenger trains, the receiving-departure yard for freight trains, the shunting yard and storing yard, the transshipment yard, the border check yard, the locomotive servicing and maintenance equipment, and the car repairing equipment, etc;some border stations shall be set with wood fumigation yard and the wheelset replacement depot, etc.

13.3.2   The number of tracks for different car yards shall be calculated and decided according to the train's pair number, nature, intensive receiving-departure degree, train marshalling plan and operation time, the effective length of the receiving-departure track shall be in line with the technical standard of the connecting track.

13.3.3 The border check yard shall be established near the national boundary line, and shall be longitudinally arranged with the receiving-departure yard, or transversely combined with the receiving-departure yard under difficult conditions. The number of receiving-departure track shall be determined according to train pair number, operation time and the through capacity, etc.

13.3.4 The passenger train receiving-departure track and passenger facilities and equipment of different track gauges shall be centralized set.

13.3.5 The locomotive, car servicing equipment and maintenance equipment for railways of different track gauges should be combined, and the publicly used maintenance track and servicing track for locomotives and cars should use three-track style.

13.3.6 The wheelset replacement workshop of the border station shall be equipped with lifting device, wheelset replacement track and wheel storing track, etc., the gauge of the wheelset replacement track should be designed as per wide gauge(narrow gauge)plus guard rail.

13.3.7 The wood fumigation yard should be built separately, far away from densely populated district, at the downwind area of prevailing wind direction and shall meet the requirement of environmental protection. The total length of the track shall be determined according to operation volume and fumigation time, etc.

13.3.8 The transshipment track of the border station should preferably select the mixed layout type of direct transshipment and landing transshipment. The length of the transshipment track should meet the requirement of whole train set transshipment. The transshipment platform should be set between the transshipment tracks of different gauges, the platform should be of the same length as the transshipment track, the platform width shall be determined according to transshipment style, operation requirement and logistic requirement, and the platform height shall be determined according to the height of the chassis base for the transshipment cars.

13.3.9 The wide gauge(narrow gauge) system and the standard gauge system shall all meet the inspection platform required by the customs and the inspection and quarantine bureau.

14 Freight Collection Station and Freight Distribution Station

14.1 General Requirements

14.1.1 The distribution of freight collection station and freight distribution station shall be determined according to railway technical policy, network planning, freight transportation layout, local or enterprise planning, etc. The freight collection station and freight distribution station of the region shall be set centralized.

14.1.2 The location of the freight collection station and freight distribution station shall comply with the following provisions:

1 Based on urban planning, plant area (mine area) planning and port planning, the freight traffic flow direction and hand-over style through the railway, the freight collection station and freight distribution station shall be set at the forward position of the junction station and wharf, near the plant area (mine area) or inside the industrial district.

2 In order for easy connection and coordination with the warehousing system, when road or conveyor belt are used for freight hauling, the location of the freight collection station and freight distribution station shall be selected after economic and technical comparison. If the mine area or freight consumption area is close and the topographical condition is suitable, they should be combined with the junction station.

14.1.3 The freight collection station and freight distribution station shall be overall layout and constructed by stages according to urban planning, mine area planning, port planning and power plant planning.

14.1.4 In the freight collection station and freight distribution station, the line-switching operation for the train with 5 000 t tractive tonnage should use mode of traction, and the line-switching operation for trains with 10 000 t or above tractive tonnage shall use mode of traction.

14.1.5 The traffic flow of the freight collection station and freight distribution station shall ensure the trains of main departure direction not to change running direction and to pass through the connection point. Under difficult conditions, the original connection mode may be retained for reconstructed stations. The relief setting of the junction station shall be comprehensively decided according to railway nature, the through capacity of the track, station throat capacity and engineering conditions.

14.2 Layout Plan of Freight Collection Station and Freight Distribution Station

14.2.1 The layout of freight collection station and freight distribution station shall be determined after comparison according to the topographical condition where the station is located, loading (unloading) technology, operation style, plant area (mine area) or wharf planning, etc., one layout plan or the combination of several layout plans may be used.

14.2.2 The layout plan of freight collection station with different loading technologies may be arranged as per Figure 14.2.2-1 and Figure 14.2.2-2.

1 If mobile loading equipment is used, the station layout should use transverse-type layout plan (Figure 14.2.2-1).

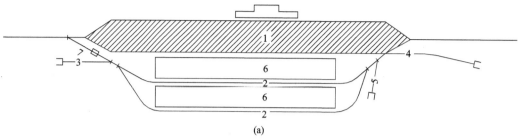

1—Receiving-departure yard of junction station; 2—Loading track doubling as receiving-departure track; 3—Safety siding; 4—Shunting neck; 5—Side repair track; 6—Loading area and stacking yard; 7—Weighing apparatus

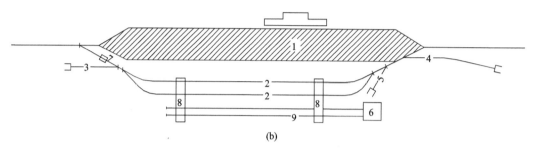

1—Receiving-departure yard of junction station; 2—Loading track doubling as receiving-departure track; 3—Safety siding; 4—Shunting neck; 5—Side repair track; 6—Junction building; 7—Weighing apparatus; 8—Mobile loading machine; 9—Running rail of loading equipment

Figure 14.2.2-1 Layout Plan of Freight Collection Station with Mobile Loading Equipment

2 If quantitative loading equipment is applied, the longitudinal type or loop type layout plan should be used for the station.

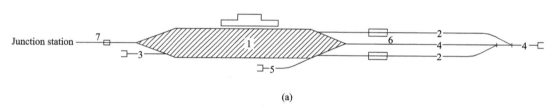

1—Receiving-departure yard; 2—Loading track; 3—Locomotive waiting track; 4—Shunting neck; 5—Side repair track; 6—Loading hopper; 7—Weighing apparatus

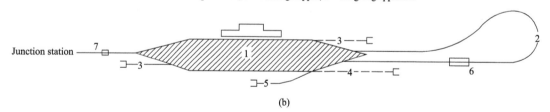

1—Receiving-departure yard; 2—Loading loop; 3—Locomotive waiting track; 4—Shunting neck; 5—Side repair track; 6—Loading hopper; 7—Weighing apparatus

Figure 14.2.2-2 Layout Plan of Freight Collection Station with Quantitative Loading Equipment

3 Mixed layout plans may be used for the station when the mine area or the port has many loading areas.

14.2.3 The layout plan of freight distribution station with different unloading technologies may be as per Figure 14.2.3-1, Figure 14.2.3-2, Figure 14.2.3-3 and Figure 14.2.3-4.

1 If the platform is provided with unloading equipment, transverse-type layout plan should be used for the station.

2 If car tippler is applied, the layout plan of station may be transverse type, longitudinal type or loop type.

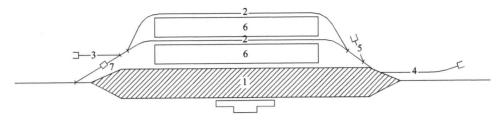

Figure 14.2.3-1 Transverse-type Layout Plan of Freight Distribution Station with Platform Unloading Equipment
1—Receiving-departure yard of junction station; 2—Unloading track doubling as receiving-departure track; 3—Safety siding;
4—Shunting neck; 5—Side repair track; 6—Unloading area and stacking yard; 7—Weighing apparatus

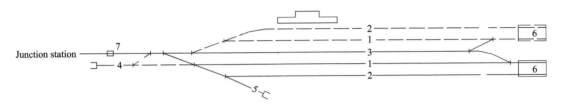

Figure 14.2.3-2 Transverse-type Layout Plan of Freight Distribution Station with Car Tippler for Unloading
1—Heavy-loaded train track; 2—Unloaded train track; 3—Locomotive running track; 4—Shunting neck;
5—Side repair track; 6—Car tippler; 7—Weighing apparatus

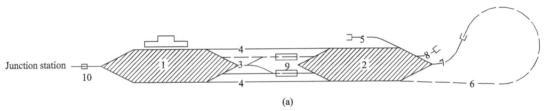

(a)

1—Receiving yard; 2—Unloading yard doubling as departure yard; 3—Locomotive running track; 4—Departure track;
5—Side repair track; 6—Departure loop; 7—Shunting neck; 8—Locomotive waiting track; 9—Car tippler;
10—Weighing apparatus

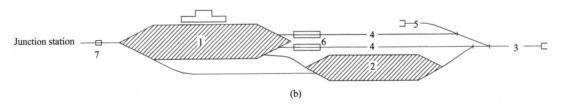

(b)

1—Receiving yard; 2—Unloaded car yard; 3—Shunting neck; 4—Unloading track; 5—Side repair track; 6—Car tippler; 7—Weighing apparatus

Figure 14.2.3-3 Longitudinal-type Layout Plan of Freight Distribution Station with Car Tippler for Unloading

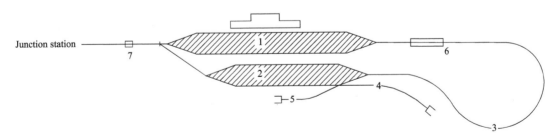

Figure 14.2.3-4 Loop-type Layout Plan of Freight Distribution Station with Car Tippler for Unloading
1—Receiving yard; 2—Departure track; 3—Unloading loop; 4—Shunting neck; 5—Side repair track; 6—Car tippler;
7—Weighing apparatus

3 If the port is with many unloading areas, the hybrid-type layout plan may be used for the station.

14.3 Main Facilities and Equipment, Number and Effective Length of Station Tracks

14.3.1 The main facilities and equipment of the freight collection station and freight distribution station include receiving-departure track, loading(unloading) track, shunting neck, loading(unloading) equipment, dust suppression equipment, anti-freezing equipment, weighing apparatus, locomotive facilities and car facilities, the equipment may be selected or other equipment can be added based on goods loading/unloading technological process.

14.3.2 The number of the receiving-departure track and the loading(unloading) track for the freight collection station and freight distribution station shall be determined according to the freight traffic volume, configuration of handling machinery, management and handover style, station operation process, time of different operation and unbalanced transportation, etc. There shall be at least two receiving-departure tracks or loading tracks for the freight collection station if they are also served as the receiving-departure track, or the receiving-departure tracks or loading tracks for the freight distribution station are also served as the receiving-departure track, the number of station tracks shall be added under the following circumstances:

1 In order to meet the need of unbalanced transportation, $1 \sim 2$ receiving-departure tracks, receiving tracks or departure tracks should be appropriately added.

2 In the case of transverse-type station or longitudinal-type station, as well as in the case of loop-type station where the shunting locomotive is used for loading(unloading) operation, one receiving-departure track shall be added to meet the requirement of locomotive running. For stations with big amount of traffic volume, it shall be determined according to specific conditions.

3 For the station with the shunting locomotive operation and the combination and disassembly operation for ten thousand tons, the number of receiving-departure tracks shall be added based on operation volume.

14.3.3 The effective length of receiving-departure track shall meet the technical standard of the railway connected to the freight collection station and freight distribution station. The effective length of the loading(unloading) track should meet the condition for whole train set loading(unloading); under difficult conditions, the effective length may be calculated and decided based on the maximum marshalling length of the train set.

15 Earthworks and Drainage of Railway Station and Yard

15.1 Earthworks in Railway Station and Yard

15.1.1 The earthworks located in railway station and yard shall be designed respectively as per the categories of earthworks of main line, earthworks of receiving-departure track, earthworks of other station tracks and yard site, etc.

15.1.2 When the shoulder elevation of earthworks in railway station/yard is controlled by flood level or tide level, the design of flood frequency or recurrence interval shall comply with the following provisions:

1 The standard for designed flood frequency of earthworks shoulder elevation for station, locomotive depot(shed), car depot(shed) and EMU depot as well as the yard site elevation of main production buildings shall adopt 1/100.

2 The standard for designed flood frequency of earthworks shoulder elevation for independent railway logistics center shall adopt 1/50.

3 The standard for designed flood frequency of yard site elevation for general production and office buildings and auxiliary housing shall adopt 1/50.

4 The designed tide level for the coastal embankment of harbour station shall adopt the high tide level with the recurrence interval 100 years. When the coastal embankment is also used as water transport terminal, the minimum tide level shall also meet the design requirement of water transport terminal.

15.1.3 The shape of earthworks surface in station shall be designed as single-slope, double-slope or sawtooth slope according to the width of earthworks surface, drainage requirement and earthworks filling and excavation. The transverse slope of the earthworks surface should not be inclined to the main line; the earthworks of main line which surrounds the car yard shall be designed as separate earthworks. In the case of receiving-departure track, the top surface of prepared subgrade as well as the top surface and bottom surface of embankment upper part or cutting's replacement layer shall be set with transverse drainage slope with gradient of not less than 2%. The drainage transverse slope for the earthworks surface of other station tracks except the receiving-departure track shall be determined according to the annual rainfall of different districts and should not be less than 2%.

15.1.4 The distance between the track center lines and the edge of earthworks surface in station shall comply with the following provisions:

1 For electrified railway and which the power(optical) cable trough is located on the shoulder, it shall be calculated and decided according to the requirement of design standard of station track, shoulder width, maintenance style, curve widening, sinking down and widening of earthworks surface, different type of power(optical) troughs of railway, OCS post and sound barrier foundation. For high-speed railway, the distance from the centerline of main line in station or the outermost receiving-departure track to the edge of earthworks surface shall not be less than 4.6 m.

2 For non-electrified railway or when various power(optical) cable trough is not located on the shoulder, the distance from the centerline of outermost track to the edge of earthworks surface shall not be less than 3.0 m; the distance shall not be less than 4.0 m if train inspection operation is needed,

and should not be less than 3.5 m when retaining wall is used under difficult conditions; for the outermost ladder track and the side where shunting operators get on and get off the trains of the shunting neck, the distance shall not be less than 3.5 m; for the car unhooking section of the hump pushing track, the distance shall not be less than 4.5 m on the side with unhooking operation, and shall not be less than 4.0 m on the other side.

3 For the single track such as connecting line and locomotive running track without passage of trains in station, the distance shall not be less than 2.8 m for soil earthworks, and shall not be less than 2.7 m for hard rock earthworks.

4 For the station track which is parallel with the main line on the same earthworks in station, such as shunting neck and catch siding, the distance from track center line to the edge of earthworks surface shall be the same with the main line in station and shall not be less than 3.5 m.

5 The minimum width of the shoulder of the outermost track in the station shall not be less than 0.6 m.

15.1.5 The design of subgrade bed for station and yard shall comply with the following provisions:

1 The earthworks standard of main line in station shall be the same as the standard of the main line in section. The earthworks standard of station approach line shall be the same as the standard of the main line with corresponding design speed.

2 If the station track is located at the same earthworks with main line, its earthworks standard shall be the same as that of the main line in station.

3 If facilities like longitudinal drainage trough and platform are constructed between station track and main line, the earthworks of station track may be built separately with the earthworks of the main line. In this case, the thickness of prepared subgrade for the receiving-departure track of high-speed railway may be 0.6 m, the thickness of embankment upper part or cutting's replacement layer may be 1.9 m, and the total thickness of the subgrade bed may be 2.5 m; the thickness of prepared earthworks for the receiving-departure track of intercity railway may be 0.5 m, the thickness of embankment upper part or cutting's replacement layer may be 1.0 m, and the total thickness of the subgrade bed may be 1.5 m; the thickness of prepared subgrade for the station tracks other than the receiving-departure tracks of high-speed railway and intercity railway may be 0.3 m, the thickness of embankment upper part or cutting's replacement layer may be 0.9 m, and the total thickness of the subgrade bed may be 1.2 m.

4 For the reconstructed section where the existing railway station is made use of, the reinforcement measures of the earthworks of main line in station shall be determined on the basis of the maximum passing speed of the train.

15.1.6 The filling material and compaction standard for station and yard earthworks shall comply with the following provisions:

1 For the main line in station, and for the station track located in the same earthworks with the main line, it shall be the same as the main line in section.

2 For the station track whose earthworks is separately set with the main line, it shall be designed as class II railway standard.

3 When facilities such as drainage ditch and platform wall are constructed on the earthworks, the backfilling of earthworks shall comply with the filling material and compaction standard of corresponding positions.

4 The standard of earth filling and compaction for passenger platform shall be designed according to the standard of the embankment upper part or cutting's replacement layer for class II railway.

5 Group D filling material may be use for the earthworks of yard site which is not in layers. The compaction standard shall comply with the relevant provisions of current national standard *Code for Design of Building Foundation* (GB 50007).

15.1.7 When different tracks are located at the same earthworks, the higher earthworks standard shall be adopted; when facilities like longitudinal drainage trough and platform are constructed between different tracks, the earthworks may be built separately with corresponding earthworks standard.

15.1.8 When railway and road are parallel with small spacing, and the elevation of road surface is higher than railway shoulder, or lower within 1.0 m, guardrail shall be set at the adjacent side of the road, and the anti-collision level shall comply with related provisions.

15.1.9 For the earthworks of receiving-departure track which cannot be separated with the earthworks of the main line as well as for the earthworks of the receiving-departure track of CWR, the transition section shall be set at the connecting part between earthworks and bridge(culvert), between earthworks and tunnel and between embankment and cutting, and the design standard shall be the same as that for the main line in section.

15.2 Drainage of Railway Station and Yard

15.2.1 The drainage system of station and yard shall be overall planned, systematically designed, and effectively connected to the local drainage system. The longitudinal and transverse drainage facilities shall be closely integrated, and the water flow routes shall be short and straight. The existing drainage facilities should be utilized for reconstructed railway station.

15.2.2 The surface water of earthworks surface for station and yard shall be drained outside of the earthworks, and the drained water shall not flush the earthworks and the slope, etc. When underground water influences the earthworks and the track bed, the drainage structures shall be constructed to drain the water into the drainage system outside the earthworks.

15.2.3 The cross section size of the drainage facilities for the station and yard shall be designed as per 50 year return period flood frequency. If sufficient basis is available, the local flood frequency may be adopted for design. The bottom width of the longitudinal and transverse drainage trough shall not be less than 0.4 m, and the depth should not be more than 1.2 m; when the depth is more than 1.2 m, the bottom width shall be appropriately enlarged.

15.2.4 The transverse gradient of earthworks surface and the maximum track number of the single slope may be determined as per Table 15.2.4. Nevertheless, various yard site for loading and unloading as well as other operations and station building site may adopt the transverse gradient of not less than 1‰.

Table 15.2.4 Transverse Gradient of Earthworks Surface and the Maximum Number of Tracks on One-way Slope

No.	Category of earthworks rock and soil	Average annual precipitation of the local area(mm)	Transverse gradient (‰)	Maximum number of tracks on one-way slope
1	Block stone, crushed stone, gravel, sandy soil(except silt), etc.	<600	2~4	4
		≥600	2~4	3
2	Other rock and soil except the above mentioned	<600	2~4	3
		≥600	2~4	2

15.2.5 The position of the longitudinal drainage facilities for station and yard shall be determined based on the shape of the earthworks surface of the station and yard. The longitudinal drainage trough within the scope of the platform should be established between the receiving-departure tracks, the receiving-departure track and the main line. Under difficult conditions, it may be established between the receiving-departure track and the platform. The transverse drainage channel of the station and yard should not cross the platform and the main line.

15.2.6 The length of the single-side drainage slope for the drainage trough of station and yard should not be more than 300 m, and transverse drainage trough may be established if necessary.

15.2.7 The gradient of longitudinal drainage facilities shall not be less than 2‰, and shall not be less than 1‰ under difficult conditions. The gradient of the transverse drainage facilities crossing the railway shall not be less than 5‰, and may be set according to specific condition under especially difficult conditions.

15.2.8 The bridge and culvert of the station should be utilized as transverse drainage facilities; when there are no bridges or culverts available, the transverse drainage trough or drainage pipe may be used.

15.2.9 The drainage facilities located at the shunting operation area, train inspection area, and loading/unloading area with operating people and vehicle passage shall meet the requirement for people and vehicle passage.

15.2.10 The inspection well or catch pit shall be set where longitudinal and transverse drainage troughs(pipes)intersect with each other, the elbow of drainage pipes and elevation changing point.

15.2.11 The drainage facilities of the station and yard should not cross with the foundation of OCS post and shelter column; they may detour under difficult conditions but the drainage capacity shall not be lowered.

15.2.12 The foundation of drainage trough in the earthworks section of frost heaving shall be 0.25 m below the frost line. When the frost line is more than 0.8 m, which cause the drainage trough excessively deep, the reinforced concrete drainage trough should be used.

16 Station Tracks

16.1 Track Design Standard

16.1.1 The type of tracks for station track shall comply with the following provisions:

1 Ballasted track should be used for receiving-departure track, and shall be used for other station tracks.

2 Ballastless track may be used for the passenger train's receiving-departure track of elevated stations, underground stations, or stations with elevated layers in platform.

3 ballastless track may be used for the receiving-departure track adjacent to the main line when the main line is ballastless track.

16.1.2 The design standard of ballasted track for station tracks shall be selected from Table 16.1.2 according to the application of the station tracks.

16.1.3 The receiving-departure track for passenger trains of extra-large passenger station and large passenger station, as well as the receiving-departure track of other stations for EMU receiving and dispatching shall be continuous welded rail.

16.1.4 The setting of stock material, guardrail, line and signal mark for station tracks shall comply with the related provisions of the current *Code for design of railway track* (TB 10082).

16.2 Rail and Auxiliary Parts

16.2.1 The same station track shall be laid with the rail of the same type; the compromise rail shall be used to connect the rails if they are of two different types.

16.2.2 When the station track is continuous welded rail, the rail of the same type as the turnout which is connected with the station track should be adopted, with cut length 100 m, and the insulated joint shall be glued insulated joint; when the station track is jointed track, the rail of cut length 25 m shall be adopted.

16.2.3 In the section of small radius curve, wearproof rail shall be used based on transportation requirement.

16.2.4 There shall be no rail joints at the following positions, and it shall be welded or glued if unavoidable.

1 Within 2 m back and front of bridge end, the temperature expansion joint of arch bridge and arch crown.

2 Within the temperature span of the steel beam equipped with rail expansion joint.

3 On the top of transverse beam of the steel beam.

4 Within the range of crossing.

16.3 Sleeper and Fastenings

16.3.1 The sleepers of different types shall not be mixed laid. When there are rail joints at the dividing point of sleepers of different types laid in sections, at least 5 pieces of sleepers of the same type shall be extended outside of the rail joints.

Table 16.1.2 Design Standard of Ballasted Track for Station Tracks

Item				Unit	Receiving-departure track								Track of rolling part of hump	Other station tracks	
					Continuous welded rail				Jointed track					High-speed railway, intercity railway	Heavy-haul railway, mixed traffic railway
					High-speed railway	Intercity railway	Mixed traffic railway		Intercity railway	Mixed traffic railway	Heavy-haul railway with heavy-loaded train	Heavy-haul railway with light-loaded train			
Rail				kg/m	60				50	60 and 50	60	50	50	50	50
Fastenings				—	II-type spring clip				I-type spring clip	II, I-type spring clip	Heavy-haul special	I-type spring clip	I-type spring clip	I-type spring clip	I-type spring clip
Sleeper	Concrete sleeper	Type		—	III	New II	III	New II	New II	New III	Heavy-haul special	New II	New II	New II	New II
		Number of sleepers laid		piece/km	1 667	1 760	1 667	1 760	1 520	1 760~1 520	1 680	1 600	1 520	1 520	1 440
	Ballast material			—	Grade I									Grade I	
	Width of top surface			m	3.4	3.3	3.4	3.3	2.9	3.0	2.9	3.1	2.9	2.9	2.9	2.9
Ballast bed	Thickness	Soil earthworks	Double layer	top ballast	cm	35	—	—	—	—	20	—	—	25	—	—
				subballast		35	—	—	—	—	20	—	—	20	—	—
			Single layer	Ballast		35	30	35	30	30	35	35	35	35	25	25
		Hard rock earthworks	Single layer	Ballast		35	30	30	30	30	25	25	25	30	25	20
		Graded crushed stone or graded sand and gravel earthworks	Single layer	Ballast		35	30	30	30	30	25	25	25	—	—	—
Side slope				—	1 : 1.75				1 : 1.5				1 : 1.5	1 : 1.5	1 : 1.5

Notes: 1 Rail refers to the new rail or reclaimed rail.
2 The receiving-departure tracks also include receiving track, departure track and marshalling-departure track, etc.
3 When the receiving-departure track of heavy-haul railway is laid with continuous welded rail(CWR), the design parameters such as the width of top surface for ballast bed and the slope shall meet the related provisions of CWR design.
4 The track of rolling part of hump refers to the section from hump crest to the outlet of retarder for shunting track.
5 Other station tracks refer to the station tracks except for the receiving-departure track and the track of rolling part of hump.
6 When the station track is maintained by large maintenance machinery, the number of sleepers shall not be less than 1 600 pieces/km.

16.3.2 The sleepers of the track connected with the turnout shall be of the same type as the turnout sleepers, otherwise the sleepers of the same type as the turnout sleeper shall be laid both ends of the turnout for transition. The transition range includes 50 sleepers respectively at both ends of the turnout on main line, and 15 sleepers respectively at both ends of the turnout on station track.

16.4 Track Bed

16.4.1 Grade I crushed stone ballast shall be used for the track bed of station track and the material shall comply with the relevant provisions of railway crushed stone ballast and the sub-ballast of railway crushed stone ballast bed.

16.4.2 The thickness of track bed for station track shall be selected from Table 16.1.2, double layer ballast shall be adopted for the track bed of the receiving-departure track for the receiving and dispatching of the trains except EMU with soil earthworks, and the track of rolling part of hump. When the annual rainfall is less than 600 mm which will not cause earthworks defect, single layer ballast may also be adopted.

16.4.3 The track bed of station tracks shall be designed as per single track, porous material shall be filled to 3 cm below bottom of sleepers between and outside of the following types of track, and the faceplate shall be laid as well:

  1 Shunting track, pushing track of hump, shunting neck, receiving-departure track and passenger train servicing siding with frequent shunting and train inspection operation.

  2 Throat area with shunting operation.

16.4.4 The shoulder width of track bed for station track shall comply with the following provisions:

  1 When the receiving-departure track is continuous welded rail, the shoulder width of the track bed shall be 0.4 m; for curve section with radius equals to or less than 600 m, the shoulder width of the track bed outside of the curve shall be increased by 0.1 m.

  2 For the shoulder of the track bed at the side of the pushing track of hump with frequent unhooking operation, the width shall not be less than 2.0 m; the other side shall not be less than 1.5 m.

  3 The shoulder width of the track bed outside of the track shall not be less than 1.5 m for shunting track, shunting neck and receiving-departure track with train inspection operation of district station and above, and passenger train servicing siding.

  4 The shoulder width of the track bed shall be 0.2 m for the track of other station tracks except the above mentioned condition, without curve widening outside of the curve.

16.4.5 The top surface of track bed for the section laid with new II and III concrete sleepers shall be parallel and level with the top surface of the middle of sleepers; the top surface of the track bed for the section laid with turnout sleepers and bridge sleepers shall be 3 cm lower than the rail bearing surface of the sleepers.

16.4.6 The track bed thickness, shoulder width and side slope of the turnout area shall be in line with the main connected tracks, inclination slope shall be done after the last turnout sleeper to eliminate the track structural height difference of the connected tracks.

16.4.7 The transition section shall be set between different track structure types such as ballastless track and ballasted track, monolithic track bed and other new under-rail foundation, or between different ballastless track structures, and the setting of the transition section shall comply with relevant provisions.

16.4.8 The special track due to requirement of operation, drainage or other demand shall be paved with monolithic track bed or other special under-rail foundation, and it shall be designed according to geological conditions.

16.5 Turnout

16.5.1 The rail type of the turnout on the main line shall be the same type as that of the main line. The rail type of the turnout on the station track shall not be lower than the rail type of the station track. When it is higher than the type of the track, the rail of the same type with the turnout or transition rail of at least 6.25 m long shall be laid each end of the turnout, or at least 4.5 m under difficult conditions, and shall not be laid continuously.

16.5.2 The selection of frog number for turnout shall comply with the following provisions:

1 The train-straight-through speed on the turnout of main line shall not be lower than the design section speed.

2 The turnout to connect the connecting track of over-line train with the main line shall be determined according to the design speed of the connecting track. When the lateral through speed of the train is between 80 km/h and 160 km/h, No. 42 single turnout may be used. When the track is jointed at the station and trains all stop there, No. 18 turnout may be used.

3 The single turnout for lateral passing train with speed between 50 km/h and 80 km/h shall not be lower than No. 18, and the single turnout with speed of not higher than 50 km/h shall not be lower than No. 12.

4 The single turnout for lateral receiving and dispatching passenger trains shall not be lower than No. 12.

5 The single turnout located at the main line for lateral receiving and dispatching freight trains shall not be lower than No. 12 at the passing station and intermediate station, and shall not be lower than No. 9 at other stations.

6 The turnout on the heavy-haul railway for lateral receiving and dispatching trains over ten thousand tons should not be lower than No. 18, and the crossover turnout on the main line and the turnout for receiving and dispatching other trains shall not be lower than No. 12.

7 Double-slip turnout shall not be adopted on the main line. When needed under difficult conditions, it shall not be lower than No. 12.

8 Double crossover shall not be adopted for trans-section continuous welded rail of main line and the section with design speed 160 km/h and above. Double crossover may be adopted when the section design speed is lower than 160 km/h under difficult conditions.

9 The turnout of the arriving(departure) terminal of the receiving-departure yard in EMU depot (point, stabling yard) should be No. 12 turnout, or No. 9 turnout under difficult conditions.

10 No. 6 symmetrical turnout shall be adopted for the rolling part of hump; other symmetrical turnouts may be retained under difficult reconstruction conditions; if it is very difficult for track connection outside of the shunting yard, No. 9 single turnout may be adopted. No. 6 symmetrical turnout may be adopted for the exit of receiving yard, rear of shunting yard, and station tracks such as logistics center and depot line.

11 The single turnout for other tracks shall not be lower than No. 9.

16.5.3 Movable frog turnout shall be used for the main line with passenger train design speed over 160 km/h.

16.5.4 The turnout with concrete turnout sleeper shall be adopted for main line and station track.

16.5.5 The type of the fastenings for turnout shall be the same as that of the major connected track.

16.5.6 The length of the inserted rail between the neighboring single turnouts f shall not be less than the provisions of Table 16.5.6-1 and Table 16.5.6-2.

Table 16.5.6-1　Minimum Length of Inserted Rail Between Two Facing Single Turnouts(m)

Turnout layout	Track category		With trains passing simultaneously at both side tracks		Without trains passing simultaneously at both side tracks
			General condition	Difficult condition	
(facing layout 1)	Main line	Train-straight-through speed $v>120$ km/h	–	–	12.5 (25.0)
		Train-straight-through speed $v\leqslant 120$ km/h	–	–	6.25 (25.0)
(facing layout 2)	Main line	Train-straight-through speed $v>160$ km/h	25.0 (50.0)	12.5 (32.0)	12.5 (25.0)
		Train-straight-through speed 160 km/h$\geqslant v>120$ km/h	12.5 (25.0)	12.5 (25.0)	12.5 (25.0)
		Train-straight-through speed $v\leqslant 120$ km/h	12.5 (25.0)	6.25 (25.0)	6.25 (25.0)
(facing layout 3)	Receiving-departure track	Passenger train	12.5 (25.0)	12.5 (12.5)	0 (12.5)
		Freight train	6.25	6.25	0
(facing layout 4)	Other station tracks	Passenger train	12.5	12.5	0
		Freight train	–	–	0

Table 16.5.6-2　Minimum Length of Inserted Rail Between Two Trailing Single Turnouts(m)

Turnout layout	Track category		Concrete sleeper turnout	
			General condition	Difficult condition
(trailing layout 1)	Main line	Train-straight-through speed $v>160$ km/h	25.0 (25.0)	12.5 (25.0)
		Train-straight-through speed 160 km/h$\geqslant v>120$ km/h	12.5 (25.0)	12.5 (25.0)
		Train-straight-through speed $v\leqslant 120$ km/h	12.5(25.0)	8.0 (25.0)
	Receiving-departure track		12.5(25.0)	8.0(12.5)
	Other station tracks	Passenger train	12.5	8.0
		Freight train	8.0	6.25
(trailing layout 2)	Receiving-departure track		12.5(25.0)	8.0(12.5)
	Other station tracks	Passenger train	12.5	8.0
		Freight train	8.0	6.25

Notes: 1　The digits in the brackets refer to the minimum length of inserted rail when No. 18 single turnouts are adopted by the track.

2　The minimum length of the inserted rail between the turnouts shall not only meet the general provisions of Table 16.5.6-1 and Table 16.5.6-2, but also shall be adjusted according to the requirement of turnout structure.

3　When the main line and station track are of continuous welded rail or for the passage of EMU trains, the minimum length of inserted rail between the turnouts shall not be less than 12.5 m.

4　When the adjacent two turnouts are of different rail type, the inserted rail shall be compromise rail.

5　When No. 6 symmetrical turnouts are adopted in continuous layout for tracks of passenger train servicing point, the length of the inserted rail shall not be less than 12.5 m.

6　Train refers to the marshalling train sets with locomotive and stated train marks, excluding the locomotive and rolling stock without complete train condition which is operated as train operation.

16.5.7　The turnout arrangement shall meet the requirement of the installation of switch transition equipment, and the transition section for ballasted track and ballastless track, jointed track and continuous welded rail.

Appendix A Structure Gauge of Railway

A.0.1 Structure gauge and basic dimension of high-speed railway shall comply with Figure A.0.1.

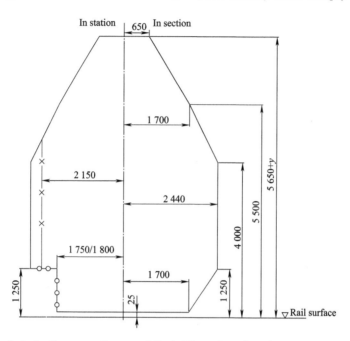

Figure A.0.1 Structure Gauge and Basic Dimension of High-speed Railway(mm)

—×—×—×— The structure gauge for signal and structural column of overhead waiting room and for OCS, overpass bridge, overbridge, electric lighting, poles and columns of shelter, etc. (not applicable to main line).

—o—o—o— ①The structure gauge of the platform(side track platform is 1 750 mm; main line platform is 1 750 mm with no trains passing by or with the passing speed not higher than 80 km/h and 1 800 mm with the passing speed higher than 80 km/h).

②The structure gauge for inverted running dwarf departure signal in station is 1 800 mm.

———————— The basic structure gauge of different buildings(structures), which is also applicable to bridges and tunnels.

y refers to the structural height of OCS.

A.0.2 Structure gauge and basic dimension of intercity railway shall comply with Figure A.0.2.

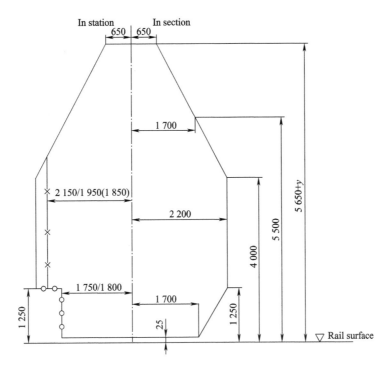

Figure A.0.2 Structure Gauge and Basic Dimension of Intercity Railway(mm)

—×—×—×— ①The structure gauge for signal and structural column of overhead waiting room and for OCS, overpass bridge, overbridge, electric lighting, poles and columns of shelter is 2 150 mm(not applicable to main line).

②The structure gauge of platform door(not applicable to main line): ground station or overhead station is 1 950 mm, and underground station is 1 850 mm.

—o—o—o— ①The structure gauge of the platform(side track platform is 1 750 mm; main line platform is 1 750 mm with no trains passing by or with the passing speed not higher than 80 km/h and 1 800 mm with the passing speed higher than 80 km/h).

②The structure gauge for inverted running dwarf departure signal in the station is 1 800 mm.

——————— The basic structure gauge of different buildings, which is also applicable to bridge and tunnel.

y refers to the structural height of OCS.

A. 0. 3 Structure gauge and basic dimension of mixed traffic railway and heavy-haul railway shall comply with Figure A. 0. 3-1~Figure A. 0. 3-3.

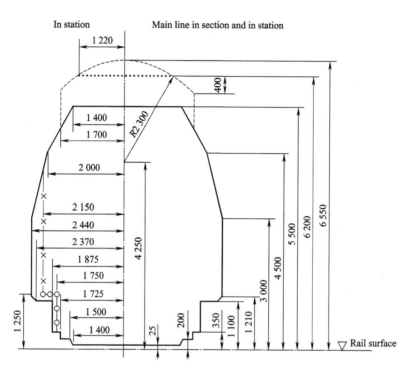

Figure A. 0. 3-1 Basic Structure Gauge and Basic Dimension of $v{\leqslant}160$ km/h Mixed Traffic Railway and Heavy-haul Railway(mm)

—×—×—×— The structure gauge for signal and structural column of overhead waiting room and for OCS, overpass bridge, overbridge, electric lighting, poles and columns of shelter, etc. (not applicable to main line).

—o—o—o— The structure gauge of platform(not applicable to main line).

——————— The basic structure gauge of different buildings(structures).

------------- Buildings(structures) which is applicable to electric traction sections, such as the overpass bridge, overbridge and shelter.

·············· The minimum height of the overpass bridge of electric traction sections in difficult conditions.

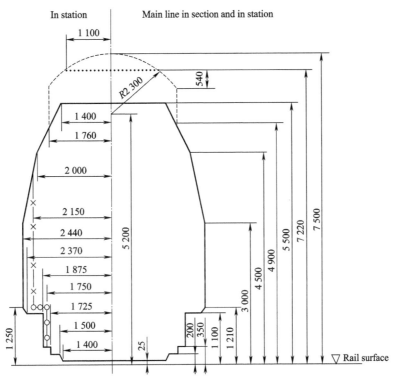

Figure A. 0. 3-2　Basic Structure Gauge and Basic Dimension of $v>160$ km/h Mixed Traffic Railway(mm)

—×—×—×— The structure gauge for signal and structural column of overhead waiting room and for OCS, overpass bridge, overbridge, electric lighting, poles and columns of shelter, etc. (not applicable to main line).

—o—o—o— The structure gauge of platform(not applicable to main line).

──────── The basic structure gauge of differenct buildings(structures).

------------- Buildings(structures)which is applicable to electric traction sections, such as the overpass bridge, overbridge and shelter.

·················· The minimum height of the overpass bridge of electric traction sections in difficult conditions.

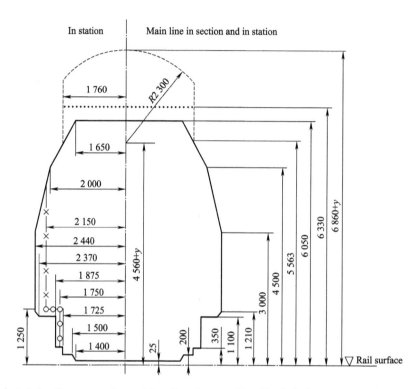

Figure A. 0. 3-3 Transportation and Loading Gauge of Double-deck Containers and Basic Structure
Gauge and Basic Dimension of Double-deck Container Railway(mm)

—×—×—×— The structure gauge for signal and structural column of overhead waiting room and for OCS, overpass bridge, overbridge, electric lighting, poles and columns of shelter, etc. (not applicable to main line).

—○—○—○— The structure gauge of platform(not applicable to main line).

──────── The basic structure gauge of double-deck container transportation which is applicable to diesel traction sections.

--------------- The basic structure gauge of double-deck container transportation which is applicable to electric traction sections.

·················· The minimum height is 6 330 mm for the wire of the contact line.

y refers to the structural height of the overhead contact system.

Appendix B Widening of Structure Gauge for Curve Section

B.0.1 In curve section, structure gauge shall be widen simply because vehicle inclination due to superelevation, The widening range of the structure gauge on the curve includes all circular curves, spiral curves and some tangents, the step widening method or curve smoothing style shall be adopted (Figure B.0.1). The widening value shall be calculated by the following formula:

$$W_1 = \frac{H}{1\,500} h \tag{B.0.1}$$

Where, W_1——widening value at curve inner side(mm);

H——height from top of rail to calculation point(mm);

h ——superelevation of outer rail(mm).

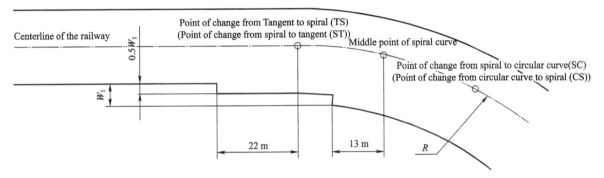

Figure B.0.1 Widening Method for Structure Gauge in Curve Section(Normal Sections)

B.0.2 The structure gauge for signal at the station side and structural column of overhead waiting room and for OCS, overpass bridge, overbridge, electric lighting, poles and columns of shelter, the structure gauge for the platform and the inverted running dwarf-form departure signal in curve sections shall be widened based on the inside and outside structure gauge of the curve. The curve widening range includes all circular curves, spiral curves and some tangents, and the widening method shall adopt the step widening method(Figure B.0.2). The widening value shall be calculated by the following formula:

$$W_1 = \frac{40\,500}{R} + \frac{H}{1\,500} h \tag{B.0.2-1}$$

$$W_2 = \frac{44\,000}{R} \tag{B.0.2-2}$$

$$W = W_1 + W_2 = \frac{84\,500}{R} + \frac{H}{1\,500} h \tag{B.0.2-3}$$

Where, W_1——curve inside widening(mm);

W_2——curve outside widening(mm);

W——sum of curve inside widening and outside widening(mm);

R——curve radius(m);

H——height of calculation point calculated from rail surface(mm);

h——outer rail superelevation(mm).

Value $\frac{H}{1\,500} h$ can be calculated by taking the inner top rail as axle, and rotating the related structure gauge angle $\theta\left(\theta=\arctan\dfrac{h}{1\,500}\right)$.

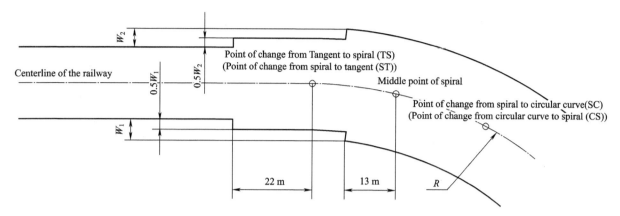

Figure B.0.2　Widening Method for Structure Gauge at Curve Sections (Pole and Column)

Words Used for Different Degrees of Strictness

In order to mark the differences in executing the requirements in this Code, words used for different degrees of strictness are explained as follows:

(1) Words denoting a very strict or mandatory requirement:

"Must" is used for affirmation; "must not" is used for negation.

(2) Words denoting a strict requirement under normal conditions:

"Shall" is used for affirmation; "shall not" is used for negation.

(3) Words denoting a permission of slight choice or an indication of the most suitable choice when conditions permit:

"Should" is used for affirmation; "should not" is used for negation.

(4) "May" is used to express the option available, sometimes with the conditional permit.